High Resolution X-ray
Diffractometry and Topography

High Resolution X-ray Diffractometry and Topography

D. KEITH BOWEN

University of Warwick

and

BRIAN K. TANNER

University of Durham

CRC Press
Taylor & Francis Group
Boca Raton London New York

CRC Press is an imprint of the
Taylor & Francis Group, an **informa** business

A TAYLOR & FRANCIS BOOK

CRC Press
Taylor & Francis Group
6000 Broken Sound Parkway NW, Suite 300
Boca Raton, FL 33487-2742

First issued in paperback 2019

ISBN-13: 978-0-85066-758-5 (hbk)
ISBN-13: 978-0-367-40063-7 (pbk)

British Library Cataloguing-in-Publication Data
A catalogue record for this book is available from the British Library

Library of Congress Cataloging-Publication-Data are available

Cover image by Kevin M. Matney, Bede Scientific Incorporated
Cover design by Youngs Design in Production
Typeset in Times 10/12pt by Best-set Typesetter Ltd, Hong Kong

Visit the Taylor & Francis Web site at
http://www.taylorandfrancis.com

and the CRC Press Web site at
http://www.crcpress.com

Contents

Preface

The purpose of this book is to provide the theoretical and practical background necessary to the study of single-crystal materials by means of high resolution X-ray diffractometry and topography. Whilst some of these techniques have been available for over fifty years, and the basic theory for even longer, it is in the last decade that they have grown enormously in importance, essentially for two technological reasons. On the one hand there has been the development of the powerful sources for scientific and industrial research: dedicated electron storage rings for synchrotron radiation, which have enabled the ideas and techniques developed in the earlier years to be applied and extended. On the other hand has been the industrial need for characterisation and control of the high-quality crystals that now form the heart of so many of the devices used for the electronics, communications and information engineering industries: integrated circuits, sensors, optoelectronic and electro-acoustic devices and microprocessors. Techniques such as high resolution X-ray diffractometry, that for half a century were rather obscure research tools, are now in daily use for industrial quality control.

Educational establishments have not yet fully responded to this growth, and the availability of trained 'characterisation' scientists and engineers is a limiting factor in many industries that manufacture devices based upon high-quality crystals. In consequence, many research or quality control laboratories are staffed by people trained in different fields, who may be unaware of both the potentialities and the pitfalls of the X-ray techniques. In this book we attempt to redress these problems. We assume only a knowledge of the materials structures and defects whose characterisation is required. We provide the elementary and advanced theory of X-ray diffraction that is required to utilise the techniques; besides covering the mathematical formalism we have taken pains to present the conclusions of the theory qualitatively and visually, to make its consequences more accessible. We discuss in depth the techniques of X-ray diffractometry and topography, including both the practical details and the application of the theory in the interpretation of the data. The limits of the techniques are explored throughout, an attitude that takes particular technological importance in the light of the present concern with very thin films and surface layers.

Preface

We are pleased to acknowledge the contributions and the stimulation of our graduate students and research associates over many years of research into materials characterisation. The content and approach of this book emerged over almost as many years of teaching advanced courses, beginning with a NATO Advanced Study Institute project in 1979 and including workshops and courses for the Denver X-ray Conference and the European Union COMETT and TEMPUS schemes. We owe special gratitude to our colleagues at Bede Scientific Instruments, where we are both directors, for funding the development of these courses and, especially, for posing interesting and difficult problems for us arising from the real needs of industrial scientists working with advanced materials. Finally, we thank the European Synchrotron Radiation Laboratory and the University of Denver for being our hosts during periods of intensive writing on sabbatical leave.

D. Keith Bowen
Department of Engineering
University of Warwick

Brian K. Tanner
Department of Physics
University of Durham

Introduction: Diffraction Studies of Crystal Perfection

In this chapter we introduce high resolution diffraction studies of materials, beginning from the response of a perfect crystal to a plane wave, namely the Bragg law and rocking curves. We compare X-rays with electrons and neutrons for materials characterisation, and we compare X-rays with other surface analytic techniques. We discuss the definition and purpose of high resolution X-ray diffraction and topographic methods. We also give the basic theory required for initial use of the techniques.

The experimental beginning of the study of materials by diffraction methods took place in 1911 when von Laue, Friedrich and Knipping[1] obtained the first X-ray diffraction patterns, on rock salt. The theoretical foundation was laid between then and 1917 by Bragg,[2] von Laue,[3] Darwin[4] and Ewald.[5] Many workers then contributed to the steady development of the subject, and we highlight in particular the work of Lang, Authier, Takagi, Taupin and Kato in developing the experimental and theoretical tools for the study of distorted and defective crystals by X-ray methods which form the main topic of this book.

In its first sixty or seventy years, two great methods dominated X-ray diffraction research. Single-crystal diffraction for structure analysis revealed the crystal and molecular structure of inorganic, organic and more recently biological compounds, and is now an essential tool in the understanding of complex processes such as enzyme action and in drug design. The powder diffraction method has become indispensable for the practical analysis of materials ranging from corrosion products to contraband. However, two developments in the early 1980s stimulated the explosive growth of high resolution X-ray diffraction, which began in the 1920s but had remained a small, specialised method. The push was the ample intensity available from synchrotron radiation sources, which allowed the rapid exploration of many new X-ray scattering techniques. The pull was the industrial requirement for rapid, non-destructive analysis of defects in highly perfect materials for the electronics industry, which was beyond the capability of the lower-resolution techniques. Interestingly, developments in X-ray optics which were enabled by the availability of large, highly perfect crystals of silicon from the electronics industry have permitted the development of laboratory-based instruments for rapid industrial materials characterisation.

The development of electron diffraction and microscopy was parallel but separate. Although there are many obvious similarities between the theories, they are usually expressed in different notations and it is common that workers trained in the one have little knowledge of the other. This is particularly unfortunate in that the techniques are almost always complementary, and many investigations will benefit by the use of both electron and X-ray (or neutron) methods. The emphasis in this book is on the X-ray methods. Electron techniques have been amply covered in a number of excellent publications;[6,7] we have not attempted to include any detailed description or assessment of them, but we shall take the opportunity to draw parallels and complementarities between the electron, X-ray and neutron beam methods.

1.1 The response of a crystal to a plane wave

Any radiation striking a material is both scattered and absorbed. Scattering is most easily approached by thinking of a plane wave. This is formally defined as one whose phase is constant over any plane normal to its direction of travel, its wavefront is a plane. It is more easily thought of as a point source of waves at an infinite distance; a perfectly collimated wave. When such a wave strikes a three-dimensional atomic lattice, each scattering point (electron or nuclear particle) acts as a source of spherical waves, whose wavefronts lie on spheres centred on the scattering points. The addition of the amplitudes of all these waves in given directions results in almost zero intensity in most directions but strong beams in some directions if the wavelengths of the wave are comparable with the spacing of the scattering centres. This is the phenomenon of diffraction.

The simplest and most useful description of crystal diffraction is still that obtained by Bragg.[2] Strong diffraction occurs when all the wavelets add up in phase. By considering an entire crystal plane as the scattering entity, rather than each individual electron, it is easy to see from Figure 1.1 that strong diffraction results when

$$n\lambda = 2d\sin\theta \tag{1.1}$$

where n is an integer representing the order of diffraction, λ is the wavelength, d the interplanar spacing of the reflecting (diffracting) plane and θ the angle of incidence and of diffraction of the radiation relative to the reflecting plane. The requirement for the angle of incidence to equal that of diffraction is not seen directly from Figure 1.1, but arises from the incorporation of scattering from many planes normal to the

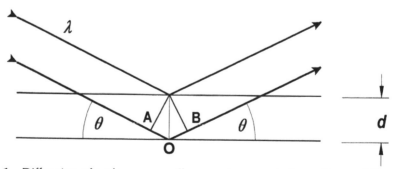

Figure 1.1 Diffraction of a plane wave off successive crystal planes. Strong diffraction results when the angles of incidence and diffraction, θ, are equal and the path difference AOB between the two beams is equal to $n\lambda$, an integral number of wavelengths. Hence the Bragg law, $n\lambda = 2d\sin\theta$

surface. A small number of planes give a very broad peak, and large numbers of planes a narrow peak, converging to a value characteristic of a thick crystal. Thus, diffraction for a given plane and wavelength does not take place over the zero angular range defined by the Bragg law, but over a small finite range. This range,

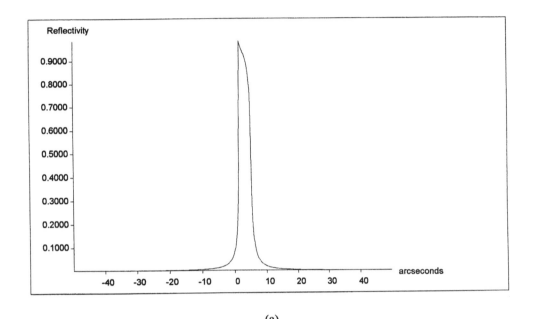

(a)

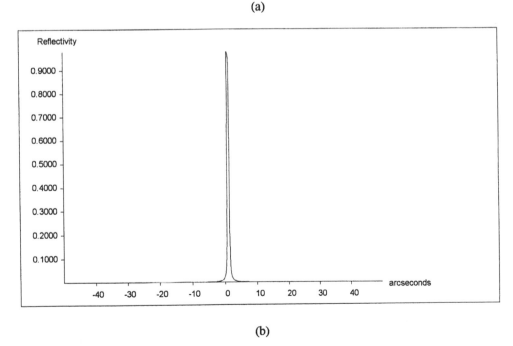

(b)

Figure 1.2 Calculated plane wave X-ray rocking curves. (a) Si 004 with CuK$_{\alpha 1}$ (0.154 nm), FWHM = 3.83 arcsec, (b) Si 333 with MoK$_{\alpha 1}$ (0.071 nm), FWHM = 0.73 arcsec, (c) Ge 111 with CuK$_{\alpha 1}$, FWHM = 16.69 arcsec, (d) GaAs 004 with CuK$_{\alpha 1}$, FWHM = 8.55 arcsec

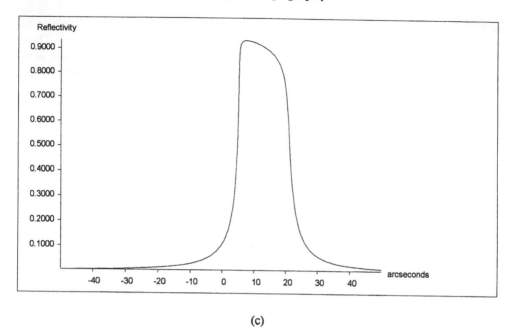

(c)

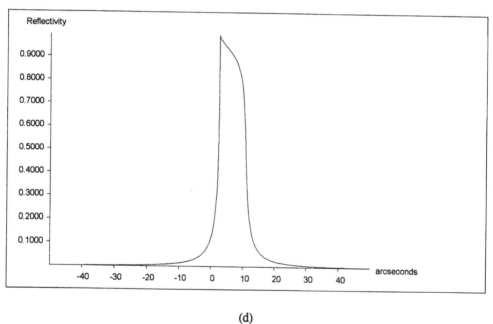

(d)

Figure 1.2 (*cont.*)

called the *rocking curve* width, varies tremendously and it governs the strain sensitivity of the technique. Examples of X-ray rocking curves for perfect crystals are shown in Figure 1.2.

Any radiation is also absorbed to a greater or lesser extent by a material. This is well described by the usual absorption equation:

$$I/I_0 = \exp\{-\mu t\} \tag{1.2}$$

where I is the transmitted and I_0 the incident intensity, μ is the linear absorption coefficient and t the specimen thickness in the direction of the X-ray beam.

The intensities of diffracted beams, or reflections as they are commonly called, depend upon the strength of the scattering that the material inflicts upon the radiation. Electrons are scattered strongly, neutrons weakly and X-rays moderately. The basic scattering unit of a crystal is its *unit cell*, and we may calculate the scattering at any angle by multiplying

- the scattering strength of an electron or nucleus
- the scattering strength of an atom
- the scattering strength of a unit cell
- the scattering strength of the total number of unit cells

all with regard to the direction of scattering and the relative phase of the scattered waves.

Electron, nuclear and atomic scattering factors are either calculated or measured and tabulated for X-rays.[8] The angular dependence of the intensity is most usefully represented by calculating the scattering strength of the unit cell for a particular lattice plane reflection, *hkl*, wavelength and, of course, crystal structure. This most important parameter is the *structure factor* F_{hkl}. This is a fundamental quantity, which appears in all the expressions for diffracted intensity, penetration depth and rocking curve width, and is calculated as follows. Waves of vector **k** scattered from two points in the unit cell which are separated by a vector **r** (see Figure 1.3), will have a phase difference

$$\exp\{-2\pi i \mathbf{k} \cdot \mathbf{r}\} = \exp\{-2\pi i (hu + kv + lw)\} \tag{1.3}$$

for the *hkl* reflection, where (uvw) are the fractional coordinates of the vector **r**. We now need to add up the waves scattered from each atom with regard to the atomic scattering factor of each atom and the phase of the wave from each atom and sum this over the unit cell:

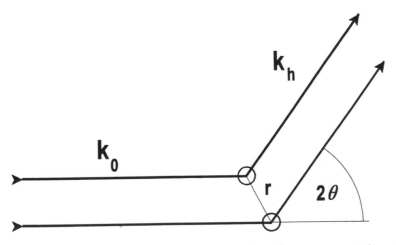

Figure 1.3 The addition of waves scattered by an angle 2θ from an atom at the origin and one at a vector **r** from the origin. The wavevectors $\mathbf{k}_0$ and $\mathbf{k}_h$ are in the directions of the incident and diffracted beams, respectively, and $|\mathbf{k}_0| = |\mathbf{k}_h| = 1/\lambda$

$$F_{hkl} = \sum_i f_i \exp\left\{-2\pi i\left(hu + kv + lw\right)\right\}$$ (1.4)

The f_i are the atomic scattering factors of the atoms of type i. These depend upon both θ and λ. The phases may cause the waves to add, as in for example Si 004, or cancel, as in for example Si 002. The larger the structure factor, the broader is the rocking curve.

There is a relationship between scattering strength and absorption strength, in that strong scatterers are also in principle strong absorbers. The parameters are often combined, as will be seen in Chapter 4, into a complex structure factor in which the real part represents the scattering factor (equation (1.4)) and the imaginary part the absorption coefficient (equation (1.2)).

1.2 Comparison of radiations for diffraction

Later chapters will deal with a more complete description of the diffraction process, but we now have enough to discuss the selection of radiations and techniques. If the structure factor and scattering strength of the radiation are high, the penetration is low and the rocking curve is broad. This is the case with electron radiation. For X-rays and even more for neutrons, the structure and absorption factors are small, penetration is high and rocking curves are narrow. These factors have three main consequences for X-rays and also for neutrons:

1 The techniques are normally non-destructive, since adequate penetration and coverage of industrial-scale samples is possible.

2 The strain sensitivity is high, since narrow rocking curves are strongly influenced by small rotations caused by strains in the crystal.

3 The spatial resolution of diffraction imaging is poor, compared with electrons, since small areas are not scattering enough radiation for good imaging, and the high strain sensitivity means that defect images are very broad.

Table 1.1 summarises the characteristics. Electron microscopy and diffraction and X-ray topography and diffraction are complementary techniques in almost every respect. The neutron techniques have applications similar to X-rays but decisive advantages in some cases, such as the study of magnetic materials and of very thick samples. The theory is well understood for all three. Two great advantages of X-ray techniques are their convenience and non-destructive nature; they

Table 1.1 Characteristics of electron, neutron and X-radiations

Characteristic	Electrons	Neutrons	X-rays
Absorption	high	very low	low
Penetration	$<1\,\mu m$	~cm	~mm
Rocking curve width	degree	<arc second	arc seconds
Strain sensitivity	10^{-3}	10^{-7}	10^{-7}
Spatial resolution	1 nm	$30\,\mu m$	$1\,\mu m$
Destructive?	yes	no	no
Cost	high	very high	medium
Convenience	good	poor	good

may thus be performed before, say, sectioning for electron microscopy and the results for both techniques combined, or they may be used for on-line quality control.

1.3 Comparison of surface analytic techniques

The penetrating power of X-rays is a major factor in their applicability to problems such as crystal growth. However, there is strong and increasing demand for methods of characterisation of very thin layers of material. Devices are now being marketed that depend for their operation on the existence of a controlled layer of material a few nanometres thick. If such layers are near the surface, as they often are for reasons of manufacture, they may also be characterised by X-ray methods using the new *grazing incidence* methods. This term is used for angles of incidence close to the critical angle for total external reflection, in practice, up to a few degrees. As seen in the previous section, the X-ray methods are particularly sensitive to strain (as are many modern devices) and it is relevant to compare the applicability of strain-sensitive surface techniques. Figure 1.4 shows a comparison of the application range

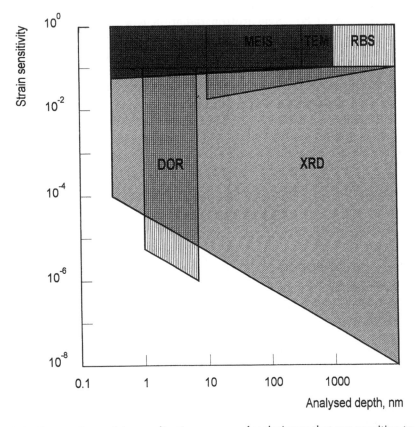

Figure 1.4 Comparison of the application ranges of techniques that are sensitive to near-surface strains. Minimum detection limits are plotted against depth resolution of the measurement. XRD: X-ray diffraction; DOR: differential optical reflectometry. RBS: Rutherford back scattering; MEIS: medium energy ion scattering; TEM: transmission electron microscopy

of the major strain-measuring techniques that are sensitive to strains in the top micrometre of crystalline material. The important role of X-ray methods is clear, even though the methods lack the spatial resolution that can be achieved with charged particle probes.

1.4   High resolution X-ray diffractometry

The term 'diffractometry' here means the measurement of the rocking curve of a sample. It is always necessary to define the incident radiation, both in wavelength and divergence, and a clear understanding of the latter is particularly important. The reference is always the plane wave rocking curve, such as would be measured with a perfectly parallel, monochromatic incident beam, and theoretical calculations are based upon this imagined radiation. Of course, any real radiation has both a wavelength spread and a divergence, and a good approximation to a plane wave curve is only found if the broadening effects of these are small compared with the width of the theoretical plane wave rocking curve. The examples in Figure 1.4 make this clear. The rocking curve widths range from 0.6″ to 12″. The divergence of a good synchrotron radiation beam could be about 1″, and that of a sealed tube X-ray source at 1 m is about 80″. Synchrotron radiation is a continuous spectrum and single-crystal rocking curves map the spectrum. The rocking curve obtained in all other cases will be dominated by the source profile, little influenced by the specimen. In other words, the instrument function for conventional powder and single-crystal diffractometers is far too great to measure subtle changes in rocking curves of nearly perfect crystals.

Better conventional collimation will not do, except for the largest synchrotron radiation installations; to obtain sub-arc-second collimation in the laboratory would require a collimator some 100 m long with a sealed-tube source, and at this distance the intensity would be impracticably low. The problem is solved by the use of a beam conditioner, which is a further diffracting system before the specimen. The measured rocking curve is then the correlation of the plane wave rocking curves of

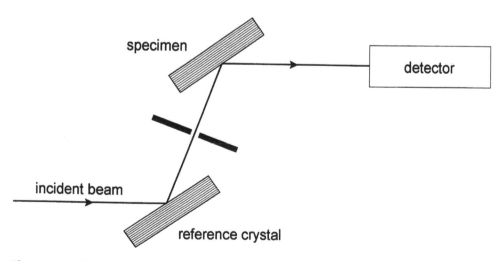

Figure 1.5   The symmetrical double crystal (+, −) setting for measuring rocking curves

the beam conditioner and the specimen crystals, from which most of the diffracting characteristics of the specimen crystal may be deduced.

The simplest conditioner is a perfect crystal of the same type as the specimen, using the same reflecting planes, with the deviation of the diffracted beam in the opposite sense to that at the specimen. This is the classic '+,– symmetrical double crystal method', as shown in Figure 1.5, which gives excellent and easily interpreted results. Many variations are, however, possible, for example to maximise the sensitivity to strain, or to emphasise the contribution of near-surface layers to the diffraction, and we shall treat these in detail in this book.

The great utility of rocking curve measurements is based upon two fundamental properties:

1   The details of the rocking curve are extremely sensitive to the strains and strain gradients in the specimen.
2   For a given structural model, the rocking curve may be computed to high accuracy using fundamental X-ray scattering theory.

In addition, the measurements are rapid and simple, and are now even used in 100% inspection for quality control of multiple-layer semiconductors. An example is shown in Figure 1.6. This is a GaAs substrate with a ternary layer and a thin cap. The mismatch between the layer and the substrate is obtained immediately from the separation between the peaks, and more subtle details may be interpreted with the aid of computer simulation of the rocking curve. This curve can be obtained in a matter of minutes. Routine analysis of such curves gives the composition of ternary epilayers, periods of superlattices and thicknesses of layers, whilst more advanced analysis can give a complete strain and composition profile as a function of depth.

1.5   Triple-axis diffractometry

Triple-axis diffractometry is an extension of high resolution diffractometry in which an analyser crystal (or crystals) is placed before the detector in order to restrict its angular acceptance. This has the effect of separating the effects of strains and tilts on the measurement, and permitting the measurement of diffuse scatter. In reciprocal space notation, which will be explained in Chapter 5, the intersection of the angular collimation of the beam conditioner crystal(s) and the analyser crystal(s) in reciprocal space defines a small volume of reciprocal space that is sampled by the detector at a given setting. Thus, complete two-dimensional reciprocal-space maps may be obtained, giving very detailed information on thin layers and surface conditions.

In later chapters we shall give both the fundamental theoretical treatment necessary to understand the complex rocking curves that arise from complex structures, and the practical experimental details required to measure them reliably and unambiguously.

1.6   X-ray topography

The essence of the topographic methods is that they map the intensity of the diffracted beam over the surface of the crystal. Defects affect the diffracted intensity,

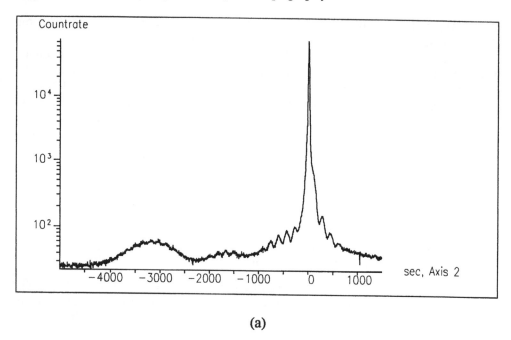

(a)

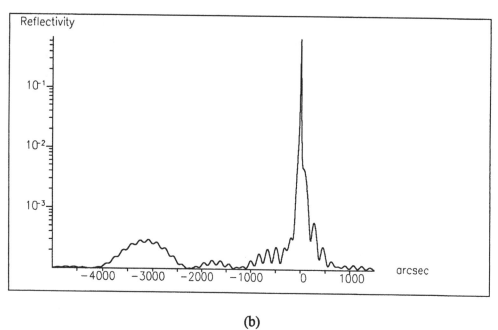

(b)

Figure 1.6 (a) An example of a measured rocking curve, for an epitaxial layer 17 nm thick of $In_{0.18}Ga_{0.82}As$, grown on GaAs, with a 0.1 μm layer of GaAs on the top. CuK$_{\alpha 1}$ radiation, GaAs single-crystal beam conditioner. (b) A theoretical simulation of this structure, by means of which the composition of the InGaAs layer was determined to <1% and its thickness to 0.1 nm

so give contrast in the image. The methods are quite sensitive enough to reveal individual dislocations, precipitates, magnetic domains and other long-range strain fields but cannot reveal point defects except in dense clusters.

1.6.1 *X-ray topography with conventional sources*

There are many methods of X-ray topography, though the most popular are the Lang method with slit-collimated radiation and the double-crystal methods, which may be thought of as high resolution diffractometry with an imaging detector. The principles of image formation and diffracted intensity are common, and may be applied to any method. The two examples show opposite extremes. Figure 1.7 shows a high resolution image of a silicon wafer, taken on an ultra-high resolution photographic plate, in which details of the strain fields of individual dislocations may be observed. Figure 1.8 shows a high-speed image taken on a television detector in which the broad features of the defect density and distribution are quickly visualised.

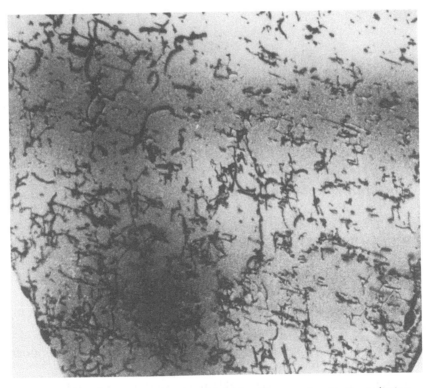

Figure 1.7 A Lang topograph (in transmission) of a silicon wafer. MoK$_{\alpha_1}$ radiation (0.07 nm), conventional X-ray tube, Ilford L4 ultra-high resolution plate. Field size 3 mm by 3.3 mm 022 reflection. Diffraction vector in direction up the page. The black lines are images of individual dislocations

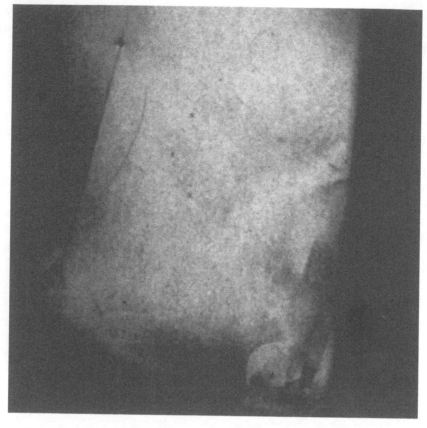

Figure 1.8 A double-crystal topograph of a GaP crystal, taken with CuK$_{\alpha 1}$ radiation from a conventional 1 kW X-ray tube. Direct imaging X-ray detector (Bede Scientific), 1 minute exposure. Cracks, sub-grain boundaries and variations in dislocation density are seen

1.6.2 *Synchrotron radiation topography*

The very high intensities and good collimations available make synchrotron radiation very suitable for topography. Against this must be set the inconvenience and slow turn-round necessitated by transport to a central synchrotron radiation laboratory. Hence, the technique is most appropriately used in the following cases:

1 For dynamic experiments, such as modelling of crystal growth or circuit processing.

2 For extensive survey topography, where the time taken on conventional machines would be prohibitive.

3 For obtaining information inaccessible to conventional methods, e.g. distorted crystals, stroboscopic topography, very thin layers, anomalous dispersion, simultaneous topography and fluorescence.

4 For rapid research into new methodologies and techniques, which may in some cases be transferred later to laboratory systems.

1.7 Summary

Crystals diffract radiation of comparable wavelength to the atomic spacing, as described by Bragg's law. The range of diffraction, or reflection, is of the order of seconds of arc for X-rays, and depends on the structure factor; this in turn depends upon the atomic species and their arrangement and indicates the strength of reflection from a given crystal plane. X-rays may probe millimetres, sometimes centimetres, into a material, or, using grazing incidence methods, penetration can be restricted to nanometres to make the techniques surface-sensitive. High (angular) resolution methods measure the reflection curve of a material to a pseudo-plane wave. This curve may be simulated using diffraction theory and a model of the material structure to determine this structure. Topographic methods give a direct image of many types of crystal defect and may be used to assess crystal perfection with high (strain) resolution.

References

1. W. FRIEDRICH, P. KNIPPING & M. VON LAUE, Proc. Bavarian Acad. Sci., 303 (1912).
2. W. L. BRAGG, Proc. Roy. Soc. A, **89**, 248 (1913).
3. M. VON LAUE, Ann. der Physik, **56**, 497 (1918).
4. C. G. DARWIN, Phil. Mag, **27**, 315 (1914).
5. P. P. EWALD, Ann. der Physik, **54**, 519 (1917).
6. P. B. HIRSCH, A. HOWIE, R. B. NICHOLSON, D. W. PASHLEY & M. J. WHELAN, Electron microscopy of thin crystals (Butterworths, London, 1965).
7. M. H. LORETTO & R. E. SMALLMAN, Defect analysis in electron microscopy (Interscience, New York, 1975).
8. International tables for X-ray crystallography, Vol. 3 (Reidel, Dordrecht, 1985).

2

High Resolution X-ray Diffraction Techniques

We first define the geometry and instrumental parameters common to high resolution diffractometry. As a reference, we then develop the duMond diagram for visualisation of X-ray optics and use it to discuss practical beam conditioners. Next we treat the principal aberrations of high resolution diffractometry: tilt, curvature and dispersion. We discuss the requirements on X-ray detectors, and finally show how to set up a high resolution measurement in practice.

2.1 Limitations of single-axis diffraction

As we have seen in Chapter 1, we need something near a plane wave in order to see the finest details of the specimen structure. A single-axis diffractometer utilises a beam that is very far from a plane wave. Thus, single-crystal rocking curves are broadened due to the beam divergence, and the spectral width of the characteristic X-ray lines.

The divergence is a function of the source size h and slit size s and the source–specimen distance a. As seen in Figure 2.1,

$$\delta\theta = \frac{h+s}{a} \tag{2.1}$$

Thus, for a typical case of $h = 0.4\,\text{mm}$, $s = 1\,\text{mm}$, $a = 500\,\text{mm}$, $\delta\theta \sim 500\,\text{arc seconds}$, far above the width of the rocking curve for highly perfect crystals, which is typically a few arc seconds.

The spectral width of the X-ray characteristic lines is approximately $\delta\lambda/\lambda \sim 10^{-4}$. This rises to 10^{-3} if both the $K_{\alpha 1}$ and $K_{\alpha 2}$ ines are diffracted by the specimen. The effect this has upon the rocking curve depends on the dispersion of the whole system of beam conditioner and specimen, and ranges from zero to very large. This will be discussed below, in section 2.6.

We therefore need to limit the divergence and wavelength spread of the beam incident upon the specimen. Beam conditioners are used to collimate and to monochromate the beam. We see that the requirement is for a system with high *angular* resolution and sufficient *monochromatisation*. We shall first discuss the characteristics of sources and the general features of high resolution diffraction systems, then analyse the requirements for beam conditioning.

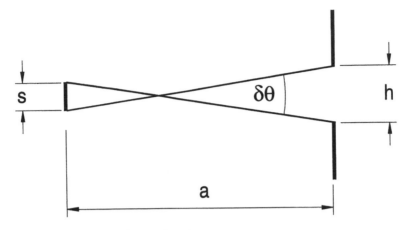

Figure 2.1 The divergence of a single-axis system

2.2 X-ray sources

Electromagnetic radiation is generated by the acceleration or deceleration of charged particles. If the acceleration is sufficiently high and the particles sufficiently energetic, the radiation is in the X-ray regime, forming a continuous spectrum. In addition, transitions of electrons between energy levels in atoms cause emission of radiation of wavelength $\lambda = hc/\Delta E$, where h is Planck's constant, c the velocity of light, and ΔE the energy difference between the levels. This gives rise to a line spectrum, whose most energetic components are in the hard X-ray regime for most elements. Both laboratory and synchrotron radiation sources are in common use for high resolution diffractometry.

2.2.1 *Laboratory generators*

In a laboratory generator, electrons are accelerated by a potential around 30 kV towards a solid target, where they are stopped by impact. The output contains the line spectrum superimposed upon a continuous spectrum. The line, or *characteristic* spectrum is characteristic of the element and is used in X-ray fluorescent analysis to identify the type and amount of an element present in a sample. The continuous radiation is also called the *Bremsstrahlung*, from the German for 'braking radiation', as it is emitted when the electrons are 'braked' by the solid. The complete spectrum from a copper target as a function of accelerating voltage is shown in Figure 2.2. It is clear that the characteristic radiation is far more intense than the continuous, and it is used almost exclusively in high resolution diffractometry.

The characteristic lines are labelled K, L, M, etc., according to the label of the electron shell to which the transition occurs, with the subscript α, β, etc. indicating where the transition started. Thus the K_α lines result from $L \Rightarrow K$ transitions, the K_β line from $M \Rightarrow K$ transitions, and so on. By far the most popular choice for high resolution diffractometry is the $CuK_{\alpha 1}$ line, because of its convenient wavelength (0.154 nm), which is useful for most interplanar spacings, and because of its high intensity. The copper target is the most easily cooled of all metallic targets, thanks

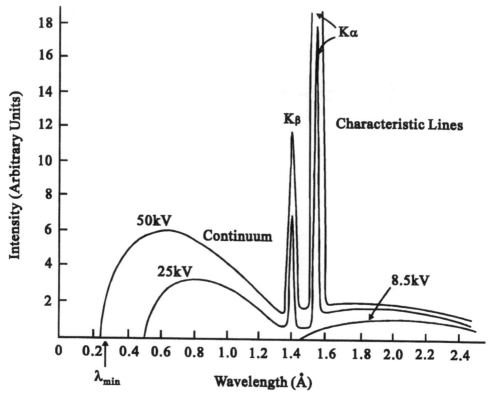

Figure 2.2   Spectrum of X-ray emission from a copper target at various excitation voltages

to its high thermal conductivity, and the output of an X-ray target is limited only by heat dissipation. The cooling may be by air, water or a combination of these with a rotating target to spread the heat more effectively. X-ray tubes run from fractions of a milliamp for a microfocus source up to about an amp for a very high power rotating target tube.

The K_α line is a doublet with separation about 10^{-3}. It is often important to remove the $K_{\alpha 2}$ line, which has approximately half the intensity of the $K_{\alpha 1}$, by the use of a beam conditioner as shown later.

2.2.2   *Synchrotron radiation sources*

Synchrotron radiation sources remove the limitation of the thermal properties of solids by confining the electron beam in an ultra-high vacuum. Magnetic fields are used to bend the beam into a closed 'ring' (in fact, a polygon with rounded corners), and the installations are called electron storage rings. Since the phenomenon was first created on earth in synchrotrons (although it had been first postulated for explaining astronomical X-ray emission), the name *synchrotron radiation* has stuck. A storage ring is simply a synchrotron run at stable rather than cyclic energy and magnetic field conditions. The installations are large, with diameters up to ~100 m, and are built on a national or international scale. The radiation spectrum is shown in Figure 2.3. All synchrotron radiation sources from simple bending magnets show this spectrum, with changes only in the vertical and horizontal position of the curve on the axes.

Since the electrons in storage rings are travelling at relativistic speeds, the emission of electromagnetic radiation is foreshortened into a cone whose axis is the instantaneous direction of motion of the electron. The radiation is therefore intrinsically collimated and is a good match to the subsequent beam conditioner. This contrasts favourably with a laboratory source, in which very little of the more-or-less isotropic emission reaches the specimen. The principal characteristics of synchrotron radiation are:

- continuous radiation spectrum
- high intensity and brightness
- high degree of polarisation
- pulsed time structure, typically at megahertz frequencies.

The sources are invaluable for the tunability of the radiation, that is where spectroscopic as well as scattering properties are important, and for experiments requiring the polarisation and time structure. However, with recent advances in X-ray tubes, beam conditioners and detectors, many scattering experiments are just as well performed in the convenience of the laboratory. Although it is difficult to attain the same intensities in the laboratory, it is in fact easier to achieve good signal-to-noise ratios. If $CuK_{\alpha1}$ is suitable for the experiment, it is likely that better productivity will be obtained with a laboratory source.

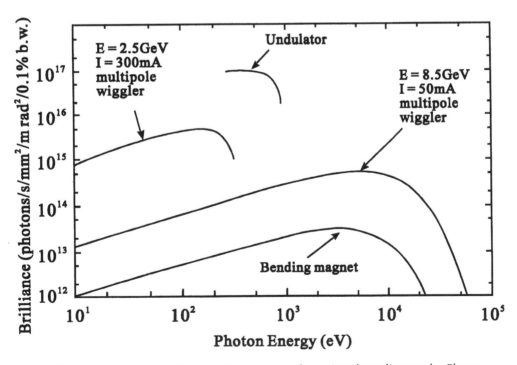

Figure 2.3 Synchrotron radiation spectrum: spectra for various beamlines at the Photon Factory

2.3   High resolution systems

Figure 2.4 shows the elements of a high resolution diffractometer. The beam conditioner controls the divergence and wavelength spread of the beam by a combination of diffracting elements and angular-limiting apertures. The latter may also control the spatial width of the beam. This falls upon a specimen, which must be adjustable in tilt so that the normal to the reflecting planes lies accurately in the plane of the diagram (assuming that the beam conditioner has been similarly set). The specimen is carried on an axis that may be adjusted to a precision of at least one arc second. The diffracted beam enters a detector, which accepts all X-rays scattered off the specimen.

The instrument shown in Figure 2.4 is a *double-axis* instrument. The first axis is the adjustment of the beam conditioner, the second is the scan of the specimen through the Bragg angle. It is irrelevant to this definition that a practical diffractometer may contain a dozen or more controlled 'axes', for example, to tune and to align the beam conditioner, to locate the specimen in the beam, to align and to scan the specimen and to control slits. It is the differential movement of the two main axes that make the measurement and determine the precision and accuracy of the instrument. This is the basic high resolution diffractometer, which is now widely used for measurements of crystal perfection, epilayer composition and thickness.

In the double-axis, high resolution diffractometer, the detector integrates the intensity scattered by the specimen over its acceptance angle. More detailed information can be obtained by collimating the detector with another X-ray optical element, at the expense of intensity. This results in the *triple-axis* instrument (Figure 2.5) in which the differential movements of the beam conditioner, specimen and analyser make the measurement and determine the precision. The triple-axis instrument is growing in popularity for research applications because of the detail that it provides, but characterisation for quality control is usually better performed on an

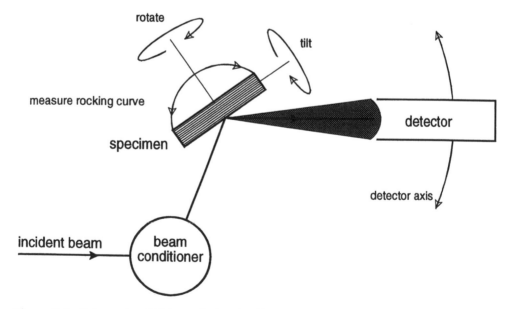

Figure 2.4   Schematic of high resolution double-axis instrument

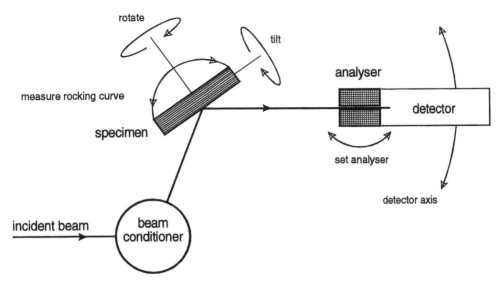

Figure 2.5 Schematic of high resolution triple-axis instrument

optically simple double-axis high resolution diffractometer because of its simpler operation, higher intensity and more rapid measurement.

X-ray topographic systems are simply high resolution diffractometers in which the beam is made large enough to cover the area of crystal required, or in which the specimen is scanned through the beam. An imaging detector is used to record the topograph. In this case sufficient collimation may be achieved by a small source and slits, as in the Lang method, but higher strain sensitivity is achieved by X-ray diffracting optical elements. In all the topographic methods, the diffracting power of the crystal is mapped across its surface.

2.4 The duMond diagram

Many types of beam conditioners have been designed, since the original double-crystal experiments of Ehrenberg, Ewald and Mark[1] in the early 1920s. These may have between one and twelve diffracting elements, different geometries, planar, circular, elliptical, paraboloidal or logarithmic surfaces. With such a bewildering variety it is essential to have some simple means of analysing their performance in order to decide between the available choices. It is logical to consider each element in turn and to see how its output is related to its input. This and the next section are primarily for reference, but are valuable in determining the choice of system and when understanding the details of performance is important.

A straightforward graphical representation invented by duMond[2] in 1937 remains extremely useful. The wavelengths and angles diffracted by a crystal must obviously satisfy Bragg's law, and duMond realised that a plot of wavelength versus incident angle would show the diffraction geometry of the crystal reflection. This is shown in Figure 2.6(a), and is obviously part of a sine curve. In effect, the crystal couples the wavelengths and the incident and scattered angles, as seen in Figure 2.6(b).

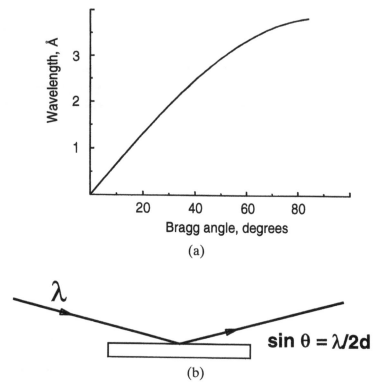

Figure 2.6 Si 220 reflection. (a) duMond diagram showing the wavelength–angle coupling imposed by the Bragg law, (b) the corresponding real-space geometry

Next we consider the input to the crystal. A plane monochromatic wave is represented by a single point on the wavelength axis, and will diffract at a single angle, as shown in Figure 2.7. The horizontal angular axis may also give information, since an aperture before the crystal will impose a certain range of angles that may be accepted by the crystal. In Figure 2.7 a wide aperture is centred on the incident beam and the crystal is at the exact Bragg angle.

However, in practice we work with a source that is polychromatic and divergent. In Figure 2.8(a) we show, superimposed on a section of the duMond diagram, the K spectral lines for copper radiation. The diagram shows the angles at which the $CuK_{\alpha 1}$ and $CuK_{\alpha 2}$ lines will diffract, and the angular aperture settings (controlled by equation (2.1)) that would be used to exclude $CuK_{\alpha 2}$ from the subsequent beam. This aperture is now only 0.06°. Figure 2.8(b) shows the real-space correspondence.

The integrated intensity of the reflection will depend also on its angular width. Therefore, the plot can be thought of as made of two lines, separated by the rocking curve width. This is a function of wavelength, Bragg angle and structure factor. On the scale of Figure 2.6 these lines are too close together to be seen, but the figure suffices to discuss a single reflection. It is now seen quite easily why single-crystal diffractometers do not have enough angular resolution to measure the fine structure of the rocking curve.

We now add another reflection. The sense of a reflection in relation to the

20

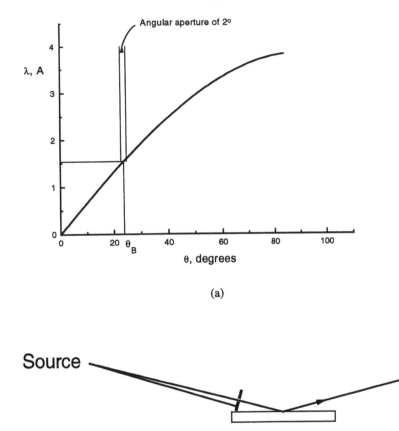

(a)

Source

(b)

Figure 2.7 Si 220 reflection with a plane incident monochromatic wave. (a) duMond diagram showing the angle at which the wave will diffract, and a 2° angular aperture that easily allows it to pass, (b) the corresponding real-space geometry

previous reflection is important. The notation used for multiple reflections is as follows:

- The first reflection is given the notation +n.
- The next reflection is given the symbol n if the planar spacing is identical to the previous reflection (same material, same plane), and m if it is different.
- This reflection is given the positive sign if it deflects the beam in the same sense as does the first crystal, the negative sign if the opposite sense.

Thus the commonly used channel-cut crystal with two reflections has the notation (+n,−n). The formal duMond diagram notation is as follows:

- The $\lambda(\theta)$ plot for the first crystal is drawn starting from the origin. This defines the direction of the incident beam and hence the zero for angular rotations of the crystals.

21

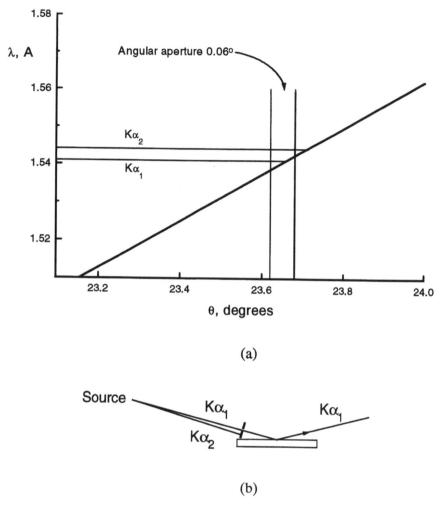

(a)

(b)

Figure 2.8 Si 220 reflection with copper-tube X-ray spectral lines. (a) duMond diagram showing the angles at which the $K_{\alpha 1}$ and $K_{\alpha 2}$ lines will diffract, and the angular aperture of 0.06° which passes only the $K_{\alpha 1}$, (b) the corresponding real-space geometry

- With the beam entering from the left and the specimen beneath the beam, the crystal diffracts when rotated anticlockwise through the Bragg angle.
- The $\lambda(\theta)$ plot for the second crystal is also drawn starting from the origin. However, there are two possibilities for sign.

  — If the second crystal has the opposite sense of deflection from the first (a +n followed by a –n or –m) then the plot goes in the same direction as that of the first crystal. The origin is the same.

    • If the second reflection is –n then the plots superimpose and no rotation is necessary for diffraction to occur from the second crystal, i.e. the crystals remain parallel. A small rotation causes the curves to separate and diffraction stops (Figure 2.9). This is also called the non-dispersive setting.

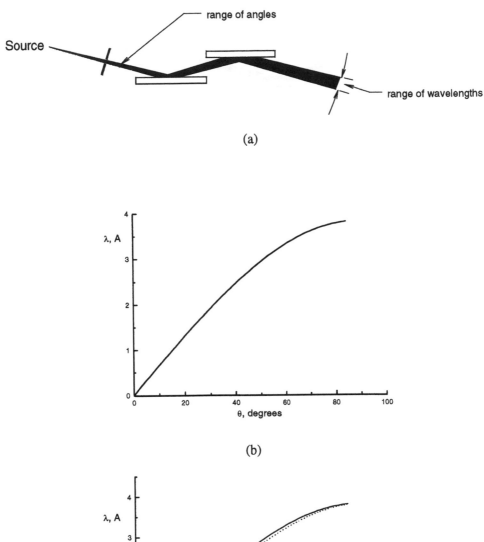

Figure 2.9 The (+n,−n) setting. (a) Real-space geometry, (b) duMond diagram when the crystal planes are parallel and the second crystal diffracts (the plots superimpose), (c) duMond diagram when the second crystal is rotated anticlockwise by $\delta\theta$ and stops diffracting. Si 220 with CuK_α

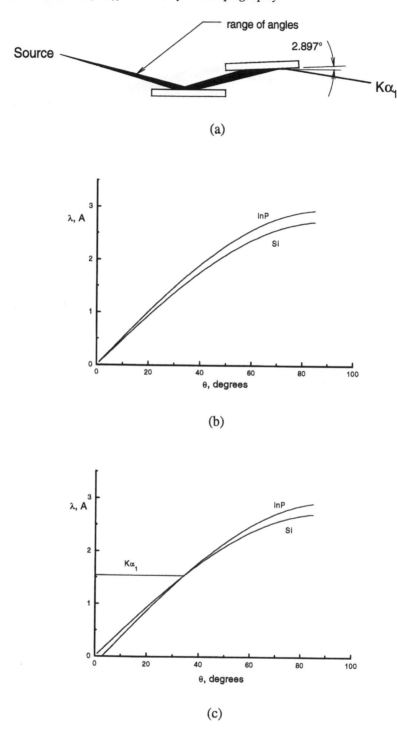

Figure 2.10 The (+*n*,–*m*) setting for InP 004 and Si 004 with CuK$_{\alpha 1}$. (a) Real-space geometry, (b) duMond diagram when the crystal planes are parallel, giving no diffraction from the second crystal, (c) duMond diagram when the second crystal is rotated anticlockwise by 2.897° and thus diffracts

- If the second reflection is −*m* then the plots diverge and a rotation is necessary for the second crystal to diffract (Figure 2.10). The reflecting range of the system comprising source, aperture and first crystal is defined as discussed above, and diffraction from the second crystal occurs when its curve on the duMond diagram overlaps with some part of this range.

— If the second crystal has the same sense of deflection from the first (a +*n* followed by a +*n* or +*m*) then the plot goes in the opposite direction as that of the first crystal. The origin is again the same. However, the second crystal will not diffract until it has been rotated by $2\theta_B$; again, the possible operating ranges of the second crystal must overlap the actual operating range of the first crystal (Figure 2.11).

- The intensity of the combined reflection is proportional to the area of overlap of the rocking curves in the duMond diagram if the source intensity is uniform across the relevant wavelength region.

- Figure 2.12 shows the relevant region of the duMond diagram for the (+*n*,−*n*) and (+*n*,+*n*) systems at much greater magnification. The shaded area is the overlap. This takes the plane wave rocking curve to be a square function, which, as will be seen in Chapter 4, is not a bad approximation.

- To allow for the profiles of both the reflecting curve and the source spectrum then we integrate the product of area element taken at constant wavelenth $A(\lambda)$ with the intensity $I(\lambda)$ at that wavelength and the reflectivity $R(\lambda,\theta)$ of each crystal, i.e.

$$I(\theta) = \int_{\lambda_1}^{\lambda_2} I(\lambda) A(\lambda) R_1(\lambda,\theta) R_2(\lambda,\theta) d\lambda \qquad (2.2)$$

The rocking curve resulting from scanning the second crystal in angle is the plot of the above intensity against angle. Mathematically this is a cross-correlation as discussed below.

The duMond diagram notation does not purport to describe quantitatively the complete intensity–wavelength–divergence transfer function of a beam conditioner. It is, rather, a highly useful qualitative and semiquantitative tool which greatly aids the visualisation and understanding of multiple diffracting elements. It has been extended to a full quantitative description by Matsushita[3] in his phase-space approach. This is of particular interest to engineers designing beam conditioners, but we do not require such detail in order to understand and to interpret high resolution diffractometry.

2.5 Beam conditioners

We shall discuss beam conditioners in some detail, since no general review applicable to high resolution diffractometry contains all the recent developments, though earlier reviews by Beaumont and Hart,[4] Kohra *et al.*[5] and Wilkins and Stevenson[6] are useful. Beam conditioners may in principle contain the following elements:

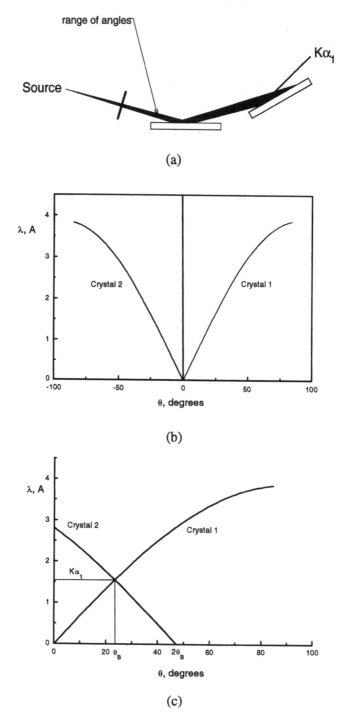

Figure 2.11 The (+n,+n) setting. (a) Real-space geometry, (b) duMond diagram when the second crystal is parallel to the first and does not diffract, (c) duMond diagram when the second crystal is rotated anticlockwise by $2\theta_B$ and the second crystal diffracts. Si 220 with CuK$_\alpha$

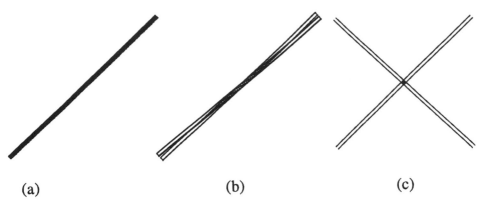

(a)   (b)   (c)

Figure 2.12 An enlarged view of duMond diagrams near the overlapping regions bounded by the aperture. (a) $(+n,-n)$ setting (the curves overlap completely), (b) $(+n,-m)$ for InP 004 and Si 004 with $CuK_{\alpha 1}$ (see also Figure 2.10), (c) $(+n,+n)$ setting aligned for maximum intensity. Si 220 with $CuK_{\alpha 1}$. The hatched areas show the band passed. Note that case (a) gives no monochromatisation, (b) gives poor monochromatisation, (c) gives excellent monochromatisation

1   apertures
2   Bragg reflections off crystals or multilayers
3   specular reflections off plane or curved surfaces.

(1) and (2) are of interest in high resolution work with laboratory sources, though specular reflections form important components of beam conditioners at synchrotron radiation sources, especially with softer radiations. Apertures are simple limitations of the lower and upper bounds of angle accepted by the beam conditioner, whose effect can be seen directly on the duMond diagram. The complexity usually lies in the design of the diffracting elements.

2.5.1   *The non-dispersive setting*

For the $(+n,-n)$ setting the duMond diagram is identical for both reflections, since the crystals and reflecting planes are identical, as seen in Figure 2.9. The region of overlap defines the wavelengths and angles for which diffraction by both crystals is satisfied. Note that when the reference and specimen crystals are parallel, all wavelengths will diffract simultaneously. Each wavelength picks out its own angle for Bragg reflection from the first crystal. If the crystals are parallel, then the same Bragg angle will be satisfied for the specimen. Thus the setting is *non-dispersive* in wavelength. A very small displacement of one crystal with respect to the other will result in no wavelength satisfying the double diffraction condition (Figure 2.9(c)).

The intensity at any angular setting resulting from the two reflections corresponds approximately to the area under the overlap of the two perfect crystal duMond diagrams, Figure 2.12. More precisely, the intensity of each 'ray' reflected by the first crystal is multiplied by the reflectivity of the second crystal for the angle at which the ray strikes the second crystal; the overall intensity is obtained by integrating over all rays (small elements of solid angle). Mathematically, this is the correlation of the two perfect crystal reflecting curves calculated from diffraction theory. If $R_1(\alpha)$ and $R_2(\alpha)$ are the reflectivities (in intensity) of the first and second crystals as functions of the angle of incidence α, then the total double crystal reflectivity $R(\beta)$ at any angle β of the first crystal relative to the second crystal is given by[7]

$$R(\beta) = \frac{\int_{-\infty}^{+\infty} R_1(\alpha) R_2(\alpha - \beta) d\alpha}{\int_{-\infty}^{+\infty} R_1(\alpha) d\alpha} \tag{2.3}$$

The denominator (normalising constant) is the integrated reflectivity of the first crystal. Figure 2.13 shows the plane wave and the double-crystal rocking curve, again for Si 220 with $CuK_{\alpha 1}$. We note the following:

- The double-crystal rocking curve is only about 40% broader than the perfect crystal reflecting curves. It can be as low as, or less than, an arc second for high-order reflections, short wavelength radiation and highly perfect crystals.
- The peak and integrated intensities of the double-crystal rocking curves are comparable with those of the plane wave reflectivity curve. Multiple-reflection beam conditioners can therefore be quite efficient if care is taken to match the divergences and acceptance angles.

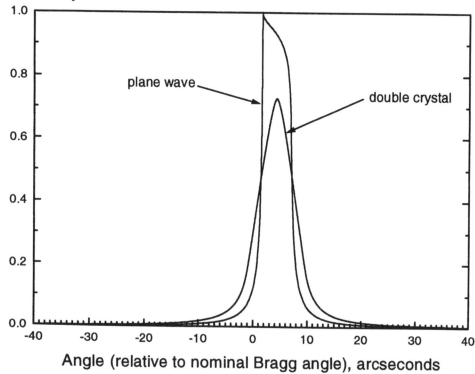

Reflectivity

Angle (relative to nominal Bragg angle), arcseconds

Figure 2.13   The plane wave and double-crystal rocking curves for Si 220 with $CuK_{\alpha 1}$

The double-crystal rocking curve is symmetric, though the plane wave reflectivity curve is not. This is a consequence of the autocorrelation, since the autocorrelation of any function is an even function.

Note that we use the characteristic lines only because they have high intensity. We can work with the *Bremsstrahlung* but the intensity is low. However, it is a way of exploiting wavelengths that do not correspond to characteristic lines.

If the second crystal is the specimen rather than a beam conditioner element, we shall have got close to the aim of measuring the plane wave reflectivity of a material. The narrow rocking curve peaks permit us to separate closely matched layer and substrate reflections and complex interference details, as already seen in Figure 1.6. The sensitivity limit depends on the thickness of the layer but for a 1 micrometre layer it is about 50 ppm in the 004 symmetric geometry with GaAs and CuK radiation. This method has been used extensively to study narrow crystal reflections since the invention of the technique.

Another way of looking at the traditional double-crystal method is that it measures the difference in reflectivity between the specimen and a reference perfect crystal. This is the original high resolution method and is still the best choice when good reference crystals are available and the specimens to be examined are of the same material. However, it does have some practical drawbacks:

1   If the specimen is changed to a different material or reflection then the reference crystal should also change to match. This is inconvenient if required very frequently, for example in a research laboratory. In a production-control environment, however, it is not important.

2   The wavelength reflected varies across the width of the beam. If the specimen is curved, the rocking curve will show separate $K_{\alpha 1}$ and $K_{\alpha 2}$ peaks, even though the arrangement is non-dispersive with plane crystals. A narrow slit is then required to eliminate $K_{\alpha 2}$.

3   The rocking curve from a single reflection has rather broad tails, as seen in the logarithmic plot of Figure 2.14. This can obscure subtle features of the rocking curve of the specimen.

Accordingly, a number of beam conditioners have been developed which both collimate and monochromate the beam, for general-purpose high resolution diffractometers.

2.5.2   *CCC design*

The problem of the broad tails of the rocking curve was attacked by Hart and Bonse, who pointed out that successive $(+n,-n)$ reflections narrow the rocking curve without much affecting its height, since the peak reflectivity is near unity (Figure 2.15). They also invented a simple method of making two parallel, perfectly aligned crystals: cut a channel out of a large single crystal with a diamond saw. This began the trend of monolithic X-ray optical elements, which now include water-cooled 'hot optics' versions for extremely powerful synchrotron radiation sources,[8] and scanning X-ray interferometers with one reflection and three transmission crystals, one scanning in angle and one in translation.[9]

These devices only work if large single crystals of low X-ray absorbance can be

log₁₀(Reflectivity)

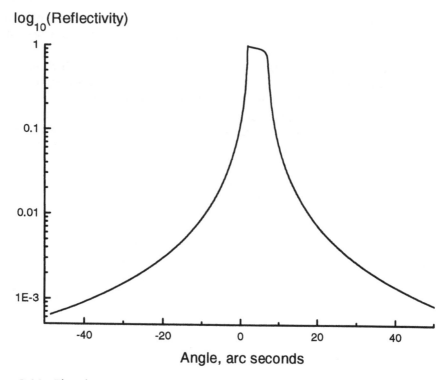

Figure 2.14 The plane wave rocking curve of Si 220 with CuK$_{\alpha 1}$, showing the tails on a logarithmic scale

obtained. Silicon is almost ideal, and is economical thanks to the electronics indus-try. It is preferable to use float-zone (FZ) material rather than Czochralski (CZ) since the latter contains some inhomogeneities from its oxygen content. Figure 2.15 shows the rocking curves produced from one, two, three and four successive reflec-tions in a channel-cut Si 220 crystal for CuK$_\alpha$ radiation, on a logarithmic scale. The effect on the tails is dramatic. Although the integrated intensity is reduced, most of the reduction takes place in the tails of the curve, where it is adding to noise rather than signal. Germanium is also available in good enough quality, though it is very much more expensive. It is more absorbing, but its broader rocking curve gives higher intensity at the expense of resolution. Diamond is used for some hot optics applications, because of its unrivalled thermal properties. Note that these curves assume ideal crystallinity, and the effects of thermal vibration and point defects mean that the curves will not drop as rapidly below a reflectivity of about 10^{-6}.

The channel-cut collimator (CCC) is not a monochromator. It is non-dispersive in wavelength and all wavelengths that hit the first crystal at their Bragg angle will reflect from the second crystal. The X-ray optics and the duMond diagram are identical to the (+n,–n) setting of the double crystal method, shown in Figure 2.9, except that the two crystals are cut out of one monolith. The spectral bandpass is determined simply by the angular aperture, controlled by a slit before or after the crystal or (for example, with some synchrotron radiation beam lines) by the natural divergence of the source.

The addition of a single monochromating crystal gives the device great flexibil-

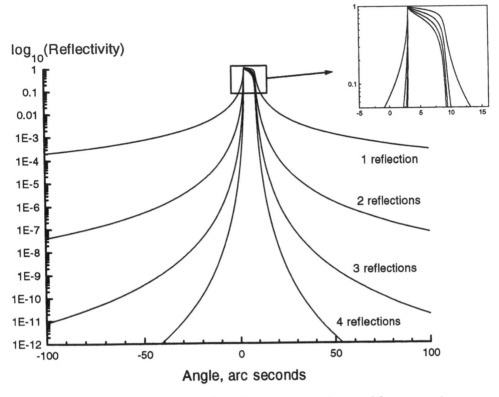

Figure 2.15 The rocking curves produced from one, two, three and four successive reflections in a channel-cut Si 220 crystal for CuK$_\alpha$ radiation. The inset shows that little intensity is lost near the peak, but the tails are greatly reduced

ity,[10] since one has independent control over the collimation, by the number of reflections, and the bandpass, controlled by the width of the reflection of the monochromating crystal. Figure 2.16 shows one variant, a three-reflection CCC to give very good angular resolution, plus a Ge 220 monochromating crystal which has a high bandpass.

2.5.3 *duMond–Hart–Bartels design*

One of the earliest designs was proposed (but rejected!) by duMond himself but only realised by Hart and popularised in diffractometry by Bartels. It comprises four reflecting crystals, in the $(+n,-n,-n,+n)$ arrangement, as shown in Figure 2.17. If the gaps between the crystals are equal, the beam is undeviated in height on exit from the conditioner. Slits between the crystal are not part of the collimation, but serve to reduce any scatter. This arrangement is convenient for diffractometer design but is difficult to adjust if four independent crystals are used. Beaumont and Hart,[4] and Bartels[11] used Hart's channel-cut crystals to make a simple design. Using germanium, the 220 reflections may be selected for high intensity, or the 440 for high resolution. Formally the duMond–Bartels design is equivalent to a CCC with three parallel reflections and a monochromating reflection, all of the same material and

Figure 2.16 The CCC with a monochromating crystal, developed by Tanner and Bowen. The four-reflection Si 220 with single Ge 200 monochromating crystal is shown, for $CuK_{\alpha 1}$

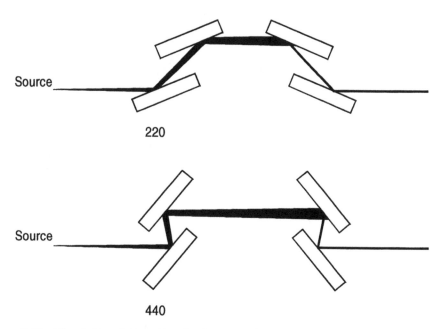

Figure 2.17 The duMond–Hart–Bartels 'four crystal' (+n,–n,–n,+n) beam conditioner design

reflecting plane. It would be possible to use different materials for the first and second monoliths to make a different compromise between intensity and angular resolution, but this loses the simplicity of the device.

2.5.4 *Asymmetric CCC and duMond*

The use of germanium rather than silicon gives higher intensity for the same geometry but has the attendant practical disadvantages of higher cost (about ten times), lower availability, greater fragility of the material and almost 50% losses from polarisation effects when the Bragg angle is near 45° as it is for the 440 reflection. Loxley, Tanner and Bowen[12] have further developed the CCC and duMond beam conditioner designs by using the principle of asymmetric diffraction. As shown in Figure 2.18, the rocking curve is broadened by an asymmetric reflection in comparison to the symmetric reflection. The reasons for this will be seen when the theory is discussed in more detail in Chapter 4. For now, we simply observe that the rocking curve width in the asymmetric case is given by

$$w_{asymmetric} = w_{symmetric} \sqrt{\frac{|\gamma_h|}{\gamma_0}} \qquad (2.4)$$

where γ_h and γ_0 are the cosines of the angles between the incident and diffracted beams, respectively, with the inward-going surface normals. For the Bragg (reflection) case, we have

$$\frac{|\gamma_h|}{\gamma_0} = \frac{\sin(\theta_B + \phi)}{\sin(\theta_B - \phi)} \qquad (2.5)$$

for glancing or grazing incidence, and

$$\frac{|\gamma_h|}{\gamma_0} = \frac{\sin(\theta_B - \phi)}{\sin(\theta_B + \phi)} \qquad (2.6)$$

for glancing or grazing exit, where ϕ is the angle between the Bragg plane and the crystal surface.

The advantage of an increased input divergence is apparent when we plot the CuK$_{\alpha 1}$ line at high magnification (Figure 2.19). Indicated on the plot are the angular widths of the symmetric and asymmetric Si 220 reflections as used for the beam conditioner shown in Figure 2.20. The symmetric reflection clearly does not take the full bandwidth of the source and thus compromises intensity further than is in fact necessary.

The Bede design is shown in Figure 2.20. The geometry and channel widths are chosen so that a simple translation of each monolith converts the device from a very high resolution beam conditioner, with the equivalent of seven parallel reflections, to a high intensity system. Moreover, the angle ϕ may be selected to give the exact degree of compromise required, rather than having it dictated by the material. It would be simple to add further slots at different values of ϕ. The value shown in Figure 2.20 is appropriate for compound semiconductors, whose rocking curves are often around 12″ wide, but a powder diffraction experiment could take the full width of the line at about 200″.

Finally we remark that proper alignment of all diffractive-optic beam conditioners normally requires that they be adjusted with a common dispersion plane. This is defined as the plane containing the incident and diffracted beam (and therefore also the normals of the diffracting plane) for any given reflection. If the dispersion planes are not parallel then some intensity will be lost from the restriction this causes to the angular acceptance in the plane perpendicular to the dispersion plane.

2.5.5 *X-ray mirrors*

With developments in ultraprecision engineering it is now possible to produce surfaces that are smooth enough to reflect X-rays efficiently at small grazing inci-

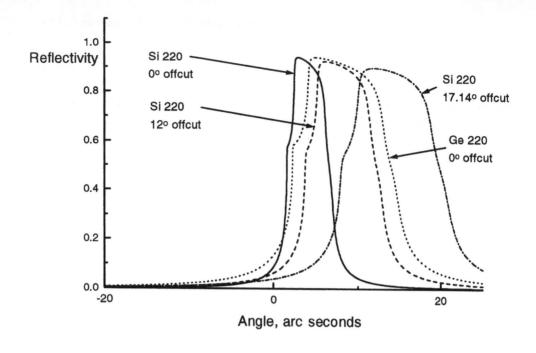

(a)

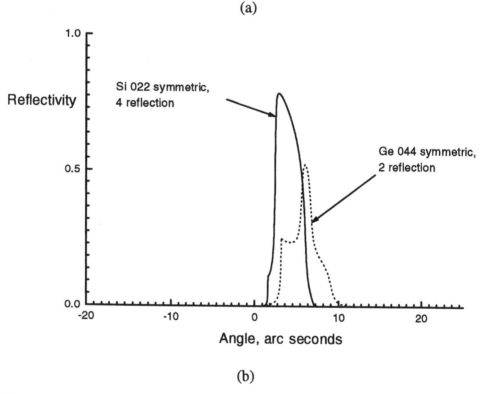

(b)

Figure 2.18 (a) The Si 220 rocking curve with different values of the offcut angle ϕ: from 0 to 17.14° in comparison with the Ge symmetric 220. The widths of the Ge 220 symmetric and the Si 022 17.14 offcut are each 12″. (b) Rocking curves for 'high resolution' settings of Si 220 (4-reflection) and Ge 440 (2-reflection). All calculations are for random polarisation; this gives rise to the inflections on the left-hand sides of the curves, and to higher polarisation losses in the Ge case

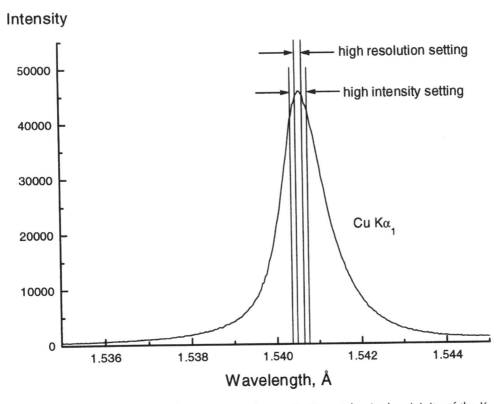

Figure 2.19 A detailed plot of the spectrum from a Cu X-ray tube, in the vicinity of the K_α lines, showing the area selected by the high resolution and high intensity settings of the beam conditioner shown in Figure 2.20

dence angles (typically <1°), and accurate enough to focus them with small aberrations.[13] These have particular applications at synchrotron radiation sources, in which the divergence is intrinsically extremely low, but beamlines are some tens of metres long and great increases in flux are achieved by collecting radiation over a larger angle. This may be achieved by curved mirrors. Doubly curved mirrors collect flux over a solid angle, and are particularly useful. However, the engineering problems are acute, requiring <1 μm figure accuracy and <2 nm surface roughness over a surface 100×500 mm for a typical advanced beamline.

The complexity of the geometry usually necessitates numerical analysis by ray tracing in order to design such beam conditioners and to predict their aberrations. The design principles are straightforward:

- An elliptical mirror is used to focus radiation from a source at one focus to the specimen at the other focus
- An ellipsoidal mirror has the same focusing properties but collects over two axes rather than one
- A parabolic mirror is used to collect radiation from one focus and deliver a parallel beam
- A paraboloidal mirror will similarly collect over two axes
- The above figures are often approximated by the simpler cylindrical or toroidal

35

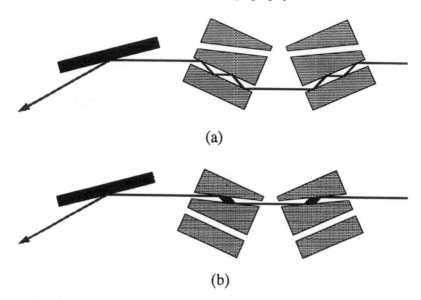

Figure 2.20 The Loxley–Tanner–Bowen combined high resolution and high intensity beam conditioner based on the duMond principle, illustrated for Si 202 with $CuK_{\alpha 1}$ radiation. (a) Geometric arrangement for high resolution, (b) geometric arrangement for high intensity

    mirrors, but these are much less efficient and give significantly increased aberrations

- Mirrors are combined with crystals to deliver monochromatic beams, but care must be taken to match divergences at the crystal with its intrinsic acceptance range (rocking curve)

High intensities can be achieved with conventional sources as well as with synchrotrons, especially if the mirror can be placed very close to a microfocus source as recently demonstrated by Bede Scientific. The same precisions and accuracies are required as for the synchrotron radiation mirrors, but the optical components are much smaller. Further improvement in efficiency with the same geometries is achievable if Bragg diffraction can be used instead of specular reflectivity. This requires that the interplanar spacing changes with position along the mirror. This is impracticable with crystals, but possible with synthetic multilayers and such beam concentrators are now becoming available in both focusing (elliptical and ellipsoidal) and collimating (parabolic and parabo-loidal) geometries. A review is given by Arndt[14] and an application of graded multilayer mirrors by Göbel.[15]

2.6 Aberrations in high resolution diffractometry

As with most instruments, erroneous results may be obtained if the high resolution diffractometer is not correctly set up. In modern instruments this is greatly assisted by the manufacturers, by prealigned components, permanent alignment on installation and by automated algorithms for specimen alignment. The primary errors to be considered are

- Tilt of the Bragg plane of the specimen, so that its dispersion plane is not parallel to that of the beam conditioner
- Curvature and mosaic structure of the specimen itself
- Wavelength dispersion arising from the bandpass of the beam conditioner

2.6.1 *Bragg plane tilt*

The theory of this aberration was worked out in the 1920s by Schwarzchild.[16] For simplicity we shall discuss the case of a beam conditioner comprising a single crystal and an aperture as in the classic double-crystal arrangement. If the Bragg planes are tilted about an axis contained in the incidence plane and the Bragg planes, then rays which are not contained in the incidence plane will not see equal angles with respect to the specimen and the reference. If we set the crystals so that the median ray (in the incidence plane) makes equal angles, then an inclined ray may make the Bragg angle for the reference crystal but will not be diffracted from the specimen (Figure 2.21). The result is that only a band of rays satisfies the Bragg conditions for both crystals. The band moves up (or down) as the crystals are rotated. The consequences are:

- the rocking curve is broadened
- the peak height is reduced
- the area under the peaks remains approximately constant
- the separation of epilayer and substrate peaks remains approximately constant

This is shown experimentally in Figure 2.22. The specimen comprises a GaAs substrate with a GaAlAs layer, and it was measured using a diffractometer with a single GaAs beam conditioner. The three-dimensional plot shows the diffracted intensity as a function of specimen rotations both parallel and perpendicular to the dispersion plane. The latter is the axis ω that we scan when measuring the rocking curve, the former, χ, is the source of aberration. The larger peak is that from the substrate, the smaller is that from the layer. In the next chapter we shall see that the composition of the GaAlAs layer may be determined from the peak separation, the substrate and layer perfections from the peak widths and the layer thickness from the integrated intensity of the layer reflection. These must therefore be measured accurately, and it is seen that aberrations are particularly acute on the peak height, but that the peak splitting is not badly affected.

The broadening δw of the rocking curve in the $(+n,-n)$ non-dispersive setting, is, to a good approximation

$$\delta w = \varepsilon\eta \tag{2.7}$$

where ε is the beam divergence normal to the incidence plane and η is the tilt between the Bragg planes of the specimen and the dispersion plane of the beam conditioner.

The broadening is purely geometric. Therefore it is more important for intrinsically narrow rocking curves than for broad ones. For strained layer systems, tilt optimisation is often wasted time. Also if you have a dispersive $(+n,-m)$ set-up, such as InP sample with a GaAs beam conditioner, the tilt broadening is usually small compared with the dispersion broadening. Again it is not then worth doing a tilt optimisation. For synchrotron radiation work, where the divergence is very low, tilt

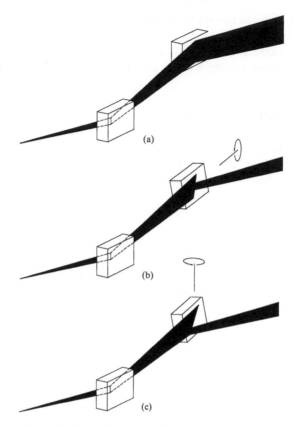

Figure 2.21 Bragg plane tilt aberration. (a) Diffracting planes parallel, diffraction occurs simultaneously over the whole height of the beam. (b) Diffracting planes skewed, diffraction only takes place over a narrow band. (c) As the crystal is rotated to measure the rocking curve, the band moves up or down the crystal. The integrated intensity remains approximately the same as in case (a) but the peak intensity decreases and the width increases

optimisation is unnecessary except in extremely high sensitivity work with silicon. However, tilt optimisation should be the norm with laboratory sources on good quality samples. For a typical collimator, ε is of the order of 10^{-2} and thus misorientation of a specimen by 2° will result in peak broadening of a large fraction of an arc minute. There are two methods for adjusting the specimen tilt systematically, considered below.

2.6.1.1 *Tilting about axis in incidence plane and Bragg plane*

From Figure 2.22 it is seen that we must alter the tilt until the rocking curve peak height is a maximum. This is possible but tedious to do manually, but it may be achieved under computer control with a suitable algorithm. The algorithm has to take account of the fact that there are multiple peaks in the field if layers are present, that the scales in Figure 2.22 are very different in the ω and χ directions, that slight inaccuracies in the construction of the instrument will skew the axis of symmetry of the (approximate) parabola in the ω–χ plane, that ultraprecision axes

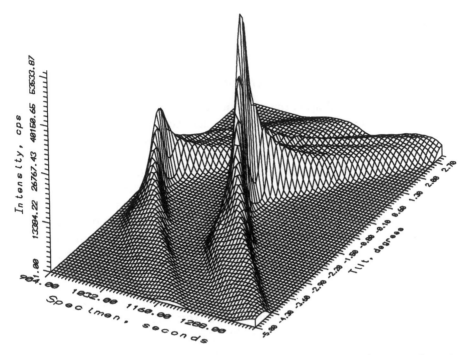

Figure 2.22   A three-dimensional plot of the diffracted intensity as a function of specimen rotations both parallel (χ) and perpendicular (ω) to the dispersion plane. The larger peak is that from the substrate GaAs and the smaller from the GaAlAs layer. CuK$_{\alpha 1}$, slit-limited from a single GaAs beam conditioner

may take an appreciable time to move and that even the best diffractometer in the world has some backlash in its axis control! An idealised peak-finding algorithm will not succeed.

Two successful automated procedures have been given. In each case one must first find the peak, and a binary search with doubled limits on each reversal[17] is the fastest search method if the peak position is unknown. One is then on either the upper or the lower branch of either the substrate or the epilayer 'parabola' (it is in fact a very shallow and distorted parabola). Fewster[18] first showed that the branch could be detected by driving down in ω until the intensity is about halved from that of the peak, then driving down in χ. If the intensity rises one is on the upper branch of the parabola, if it falls one is on the lower branch. Fewster then alternated steps in ω and steps in χ until the peak was found. The latter is determined by the intensity falling instead of rising on beginning the χ step.

This algorithm works in most circumstances but it is slow, and the end condition is not unique. The I–ω–χ field narrows sharply near the optimum position, so that at certain intensity levels the intensity contour is double-valued in χ; that is, it is shaped like the Greek letter Σ rather than a simple parabola. With some starting conditions the peak will be found on one of the two left-hand points of the Σ rather than the correct right-hand point. An algorithm been developed by Loxley *et al.*[17] which avoids this problem. Fewster's condition is used to determine the location of the branch of the parabola. The axis χ is moved in the known direction until the intensity has dropped by about 10%, then a step in ω is made until the intensity has peaked and then dropped by 10% from the peak. Iteration of this procedure sends the specimen adjustment towards the optimum, while ensuring that it stays on the ridge of the surface and does not encounter the ambiguous conditions.

2.6.1.2   *Rotation about the specimen surface normal*

A different method, based upon the rotation α of the specimen about the surface normal, has been described by Tanner, Chu and Bowen;[19] this depends upon a geometrical location of the end point and should in general be more rapid for a comparable accuracy. It also lends itself to automation. However, it will only work for symmetrical reflections. It uses the concept that the crystal's reflecting planes are parallel when the reflecting plane normal is rotated to lie in the incidence plane, and the Bragg angle has been found. The principle is illustrated in Figure 2.23, which shows a 360° α-scan of a sample with a single epilayer, indicating the position of α that gives perfect alignment of the diffracting planes. The plot is symmetrical about the ideal position, irrespective of the complexity of the rocking curve, thus giving a geometrical means of finding the ideal position without having to find any maximum positions or even having to decide which is the substrate peak.

The peak is first found by a positive sense search on the α axis. Since wafers are usually some fractions of a degree misoriented from the nominal crystallographic plane – and sometimes deliberately offset by up to 4 degrees – this provides a rapid scan over a relatively large angle. If this fails, a binary search is made on the ω axis. When the Bragg position is located on the positive sense α-scan, this axis is searched in the opposite direction to find the 'conjugate' Bragg peak. The correct setting of α is midway between these two peaks.

All automated algorithms require extensive tests and traps to detect unusual and fault conditions, such as failure of the X-ray source, grossly misaligned samples, and even perfectly aligned samples!

2.6.2   *Curvature and mosaic structure*

If the specimen crystal is curved, there will be a range of positions where the diffraction conditions are satisfied even for a plane wave. The rocking curve is broadened. It is simple to reduce the effect of curvature by reducing the collimator aperture. For semiconductor crystals it is good practice never to run rocking curves with a collimator size above 1 mm, and 0.5 mm is preferable. Curved specimens are common; if a mismatched epilayer forms coherently on a substrate, then the substrate will bow to reduce the elastic strain. The effect is geometric and independent of the diffraction geometry. Table 2.1 illustrates this effect.

A mosaic spread in the crystal will give a similar broadening. If the beam can be reduced to below the average mosaic cell size then the rocking curve will narrow. For a broader beam, substrate quality may be assessed by measurement of the full width at half height maximum (FWHM) of the high resolution Bragg peak, provided that the tilt and curvature aberrations have been removed as described above. The increased width of the rocking curve then arises principally from the variation in tilts and dilations in the crystal from its dislocation structure. It is worth noting, however, that when crystals contain sub-grains of size comparable to the beam area, the rocking curve FWHM can be a strong function of beam size. Measurements made with a different size or divergence beam may well yield differ-

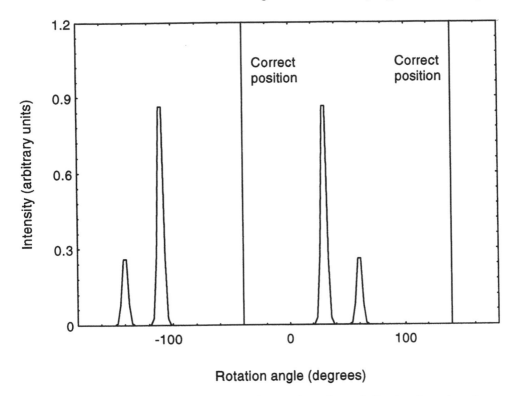

Figure 2.23 A 360° α-scan of a sample with a single epilayer, indicating the value of α that gives perfect alignment of the diffracting planes

Table 2.1 Broadening of rocking curve from 004 reflection from InP. CuK_α radiation with a beam diameter of 0.5 mm

Radius (m)	Substrate peak width (arc seconds)
Infinity	9.8
100	10.1
10	14.1
5	22.6
1	104

ent values for the FWHM if there is a sub-grain structure and this can be the cause of protracted and frustrating arguments between substrate vendors and customers! The triple-axis reciprocal space mapping methods are preferable in such cases, since they separate the effects of tilt (mosaic spread) and strain.

2.6.3 *Dispersion*

If the beam conditioner delivers a significant spread of wavelengths, then the rocking curve may be broadened. Whether it is or not depends upon the diffraction geometries. The $(+n, -n)$ geometry is non-dispersive and all wavelengths diffract at the same relative setting of specimen and reference. In the $(+n,-m)$ cases, in which the lattice plane spacings of reference and specimen are not equal, the rocking curve is broadened by dispersion. The broadening $\delta\theta$ in these cases is given by

$$\delta\theta = \frac{\delta\lambda}{\lambda}\left|\tan\theta_1 - \tan\theta_2\right| \qquad (2.8)$$

where θ_1 and θ_2 are the Bragg angles of reference and specimen crystals and $\delta\lambda/\lambda$ is the bandwidth of the X-ray wavelengths passed by the beam conditioner. For the $(+n,-m)$ setting this can be taken as the separation between the characteristic K_α lines as these lines contribute most to the intensity. The separation is the difference in the $K_{\alpha1}$ and $K_{\alpha2}$ angular separations in the reference and specimen reflections. For example, if InP is measured using a single GaAs crystal as beam conditioner, there is a broadening of the order of 25 arc seconds. This is adequate for many purposes, including the study of poor-quality materials and also strained layer systems, where the layer peaks are typically 1000 seconds or more from the substrate. It would be inadequate for quality analysis of good substrates.

The dispersive $(+n,-m)$ mode has already been seen clearly with the duMond diagrams, Figure 2.10. Here, the curves are no longer identical and the crystals must be displaced from the parallel position in order to get simultaneous diffraction. As the crystals are displaced, so the band of intersection moves up and down the curve. When the curves become very different, the $K_{\alpha1}$ and $K_{\alpha2}$ intensities are traced out separately. Then the peaks are resolved in the rocking curve, and if no better beam conditioner is available it is important in such cases to remove the $K_{\alpha2}$ component with a slit placed after the beam conditioner. A slit placed in front of the detector, with the detector driven at twice the angular speed of the specimen, also works very well. This is in effect a low resolution triple-axis measurement.

In the general multiple-crystal beam conditioner case, there is no universal formula for broadening. Rather, the duMond diagram is constructed for the beam conditioner and the shape of the passed band in $\delta\lambda$ and $\delta\theta$ is determined. The specimen crystal is then represented on the duMond diagram and scanned through the beam conditioner bandpass. For accurate simulation, of course, the true shape of the rocking curves must be taken into account, since they are not simple rectangular functions. The duMond diagram representation is most useful for obtaining a semi-quantitative understanding of the properties of a given beam conditioner. For example, if the bandpass region is very much smaller in both directions than the rocking curve width of the specimen, the latter will be accurately measured though the intensity may be small. The assumption that the broadening in a multiple-reflection monochromator is equal to the divergence of the exit beam is usually adequate.

We can now answer the common question, is it better to use the double crystal or the four-reflection monochromator for high resolution work? We see from the above, that if the first and second crystals are well matched in Bragg angle, the resolution loss in double-crystal will be simply that from the correlation of equation (2.3). This is about 1.4 times the plane wave rocking curve width. If they are not well matched, then the further loss of equation (2.8) arises. This is plotted in comparison with the high resolution and high intensity settings of the Loxley–Tanner–Bowen monochromator (Figure 2.20) in Figure 2.24. The diagram is drawn for a Si 004 first crystal; for another material, the 'V' would be a very similar shape, but centred on the Bragg angle of the material. The minimum value of about 2 arc seconds in the

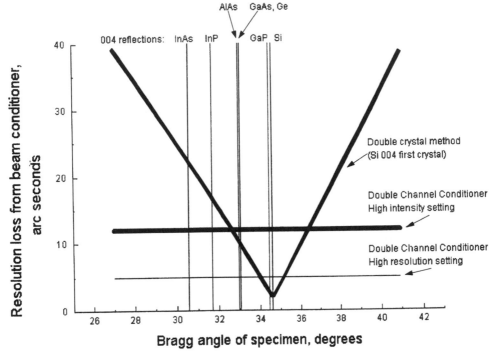

Figure 2.24 The angular resolution of a high resolution system, comparing the 'performance functions' of the double-crystal method with a four-reflection monochromator (Loxley–Tanner–Bowen, high resolution and high intensity settings). The performance functions show the resolution (ordinate) and intensity (width) of the conditioner, read off at the intersection with the vertical line at the Bragg angle of the specimen crystal. The double-crystal setting (using Si 004 as an example) is superior in both resolution and intensity when the reference and sample crystals are well matched, but the monochromator systems are superior if a wide variety of samples are to be measured

double-crystal arrangement is the difference between the width of the resulting rocking curve and that which would be obtained for a perfect plane wave. This value would increase to about 5 seconds for the heavier III–V materials such as GaAs. However, if the crystals are not well matched the four-reflection monochromators are superior and hence more valuable for general work.

2.7 Detectors

The detector must obviously have good sensitivity to X-rays, a good dynamic range and a low background noise. The choice for high resolution measurements has traditionally been the ionisation chamber, scintillation counter or proportional counter with solid-state detectors used occasionally for low-noise applications. It is necessary for good noise performance to count single photons rather than to measure a current or voltage from the detector. This eliminates ionisation chambers except for use on synchrotron radiation sources, where their very high saturation level can be exploited and low-noise performance is less essential. We summarise the behaviour of counting detectors here; a detailed treatment may be found in Knoll.[20]

2.7.1 *Proportional counters*

A proportional counter consists of a tube filled with a gas such as xenon, with positive and negative electrodes. The negative electrode is a thin wire maintained at a potential around −2 kV. Incoming photons ionise gas molecules. These drift towards the negative electrode, until the field enhancement around the thin wire is sufficient to multiply them by the cascade effect, and cause a charge pulse on the wire. The pulse is quenched by the addition of a quench gas, normally a halogen or hydrocarbon which reacts with the ions and stops the cascade.

The proportional counter can give very good low-noise performance, well below 1 cps. It also has the property that the charge pulse is proportional to the incoming photon energy, with a resolution around 10% $\delta E/E$, so a pulse height analyser may be used to clean up the signal against spurious photons, for example cosmic rays. Its main disadvantages are that it is difficult to quench the pulse rapidly enough to get a count rate above about 10^6 cps, and that the quench gas is used up, so the detector has a limited lifetime, typically two years or so. The efficiency is rarely above 50%.

2.7.2 *Scintillation counters*

The scintillation counter works by first converting the X-ray photon to optical photons via a scintillator. The optical pulse is then amplified by a photomultiplier. Although these are only about 5% efficient, several hundred optical photons are produced by one X-ray photon so the quantum efficiency approaches 100%. Originally, scintillation counters had poorer noise and a lower dynamic range than proportional counters, but recent developments have reversed this position. Using fast scintillators with a 10 ns pulse width, photomultipliers with low-radioactivity glass and high stability electronics, a dynamic range of 0.1 to 3×10^6 cps has been achieved at high quantum efficiency[21] (Figure 2.25). The energy resolution is not so good as the proportional counter, at around 50%, but is still good enough to discriminate against the cosmic ray background.

2.7.3 *Solid-state detectors*

In solid-state detectors the incoming photon illuminates a piece of silicon or germanium, creating electron–hole pairs. These are separated by a biased p-n junction and amplified with a charge amplifier. The pulses have to be measured with a multichannel analyser for spectroscopy, but for diffraction a single-channel analyser (pulse-height analyser) suffices. The noise performance is very good, below 0.01 cps, but the upper limit is rather low for diffraction at a few times 10^4 cps. Solid-state detectors have the highest energy resolution, around 2%. They are therefore mainly used for experiments in which the count rate is very low and good signal-to-noise is

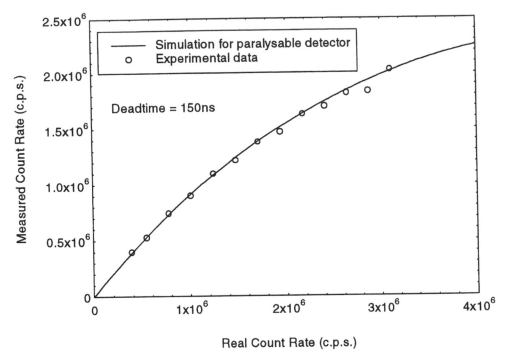

Figure 2.25 The behaviour of an advanced scintillation detector at high count rates. After Cockerton and Tanner[21]

required. However, the preamplifier has to work at low temperature, usually under liquid nitrogen, for such low noise to be achieved.

Recent advances in semiconductor materials have made feasible room-temperature solid-state detectors made from crystals such as mercuric iodide. At present they are not competitive in noise and dynamic range with advanced scintillation or proportional detectors, but may become so in future.

2.7.4 *Position-sensitive detectors*

For normal high-resolution studies, the detector has a fairly large aperture, typically 5–25 mm. There are some applications for position-sensitive detectors for low resolution triple-axis diffractometry. These are in essence gas-filled proportional counters, with detection of the position of the charge pulse along the wire. This is achieved by comparing the times taken for the pulse to arrive at either end of the wire. Again, these are limited to about 10^4 cps maximum count rate, but the position resolution can be a few tens of micrometres.

2.7.5 *Dead time correction*

At the upper end of the range, counting detectors saturate owing to pile-up of the pulses from the ion cascade, scintillator or silicon crystal. These pulses have well-defined characteristics, and a statistical treatment allows the count rate to be corrected for the 'dead time' during which a detector cannot handle a new pulse because it is still processing the last one. There are two limits of behaviour: the paralysable detector, in which the dead time switches the detector completely off and the accumulation of successive pulses eventually reduces the quantum efficiency to zero; and the non-paralysable detector which simply gets less and less efficient as the intensity increases. The equations describing the limiting cases are:

Paralysable:

$$R_{measured} = R_{true} \exp(-R_{true}\tau) \tag{2.9}$$

Non-paralysable:

$$R_{measured} = R_{true}/(1 + R_{true}\tau) \tag{2.10}$$

where $R_{measured}$ and R_{true} are the measured and true count rates respectively and τ is the dead time during which the detector cannot sense a new pulse. A practical detector is likely to be somewhere in between. The most useful approach may therefore be to measure curves such as that illustrated in Figure 2.25 and fit them to a cubic spline polynomial rather than one of these simple equations. An improvement factor of about five in the dynamic range can thus be achieved.

2.8 Setting up a high resolution experiment

The requirements for full alignment and specimen set-up are given in this section. The way in which these are achieved will differ between instruments and manufacturers, and some of them may well be built into the instrument in the factory. Even so, it is reasonable for the user to enquire how, and to what precision, these are achieved.

The symmetric setting, illustrated in, for example, Figure 2.4, is the commonest and easiest diffraction experiment, since the reflecting planes are parallel to the specimen surface – normally 004 reflections in semiconductor wafers. This gives information about the strains normal to the surface but it is important to note that this tells us nothing about the deformation of the layer in the plane of the substrate. No information on layer coherency or relaxation is available in the symmetric geometry. By using Bragg planes which are not parallel to the surface, we obtain a measure of the in-plane strains and hence may measure relaxation. This is the asymmetric setting, for example the 224 or 113 reflections from an (001) wafer. Both symmetric and asymmetric settings must be available for complete characterisation.

2.8.1 *Instrument alignment*

1 Determine the take-off angle from the X-ray tube. This controls the beam size, and a resultant beam size of approximately 0.5 × 0.5 mm is about right. Use the maximum take-off angle consistent with this size. Align the instrument so that the beam conditioner points at the source at this take-off angle.

2 The beam conditioner may have some angular, tilt and translation adjustments. These should be set so that the beam delivered

- passes along the axis of the instrument, to a precision , 0.1°
- passes over the specimen (ω) axis, to a precision ~0.1 mm
- is parallel to the diffractometer body (perpendicular to the ω axis), to a precision ~0.1°
- is maximised in intensity
- contains only the $K_{\alpha 1}$ line

3 Clamp the diffractometer if appropriate so that these alignments are preserved.

2.8.2 *Specimen alignment and peak finding*

1 Mount the specimen on the holder, so that it is securely held without vibration or creep but is not strained by the mounting. This is achieved by fixing the specimen at one point only, using PVC adhesive tape (as supplied for crystal growth laboratories). Kinematic location (resting under gravity with minimum constraints) may be used in some diffractometers. At all costs avoid over-constraint of the specimen.

2 Ensure that the point to be measured on the specimen is that being hit by the beam. This is best achieved by aligning the specimen so that its surface lies on the ω axis. An adjustment perpendicular to the specimen surface is required for this alignment, which should be made to 0.1 mm for normal work, and 0.01 mm or better for grazing incidence work.

3 For symmetric reflections the peak search may now begin. For asymmetric reflections, the specimen must be rotated about its normal until the desired diffraction vector lies in the incidence plane of the beam conditioner. This is normally the diffractometer surface. An accurate knowledge of the orientation of the specimen in two axes is required to set asymmetric reflections; this is usually taken from the position of the orientation flat or groove.

4 Switch on X-rays and search for a peak by scanning the ω axis. The binary peak search described in section 2.6.1.1 is the most efficient if the position is unknown. An intense peak is easily seen, therefore the scan can be fast, with short detector counting times.

5 When a peak is found, do a rapid scan in its region to find the narrowest peak. This is the substrate peak and is normally also the most intense.

6 Use either of the methods described in sections 2.6.1.1 and 2.6.1.2 to optimise the tilt setting of the specimen.

2.8.3 *Measuring the rocking curve*

Now the specimen is perfectly aligned and it remains to measure the rocking curve. This is normally done by selecting a step size on the ω axis and a counting time per step for the detector. Some systems will allow instead the selection of a total number of counts at each step, so that constant statistical accuracy is achieved. In either case the diffracted intensity is found as a function of the angle on the ω axis.

The step size and counting time must be chosen so as to record all the significant features of the specimen for the measurement. If only the peak splitting is needed, for example for a composition map on a ternary wafer, the measurement may be made very rapidly and data processing used to obtain good locations of the peaks. If fine features are to be extracted, such as weak interference fringes, the data must be collected more slowly. The most important principle to grasp is that the emission of X-rays follows a Poisson distribution. Therefore the standard deviation σ of a number of counts is equal to the square root of the number of counts itself, to a very good approximation. Note that it is the total number of counts, not the count rate, that is important. A good rule is that the acceptable error in a reading is $\pm 3\sigma$. Thus the precision of a count rate of 100 counts per second, measured with a counting time of 5 seconds, is $\pm 3\sqrt{(100 \times 5)} \approx \pm 67$ counts or ± 13 counts per second.

There is no point in measuring for longer than is required, nor is there ever any point in collecting bad data. As will be discussed later, an important measurement strategy in materials characterisation is to simulate the expected rocking curve using a dynamical diffraction computer package (see Chapter 5), and, noting the intensity of the main peak, decide what counting time is required to distinguish the features of interest. The properties of the diffractometer, in particular the intensity and bandpass of the beam conditioner, the resolution of the $\bar{\omega}$ axis and the saturation and noise levels of the detector, will determine the scale of the finest features that can be resolved.

2.9 Summary

In this chapter we have discussed the choice of X-ray sources and shown the appropriate regimes for sealed tubes, rotating anodes and synchrotron sources. We have treated the valuable duMond diagram for qualitative design of X-ray optics, and applied it to various high resolution beam conditioners. We have covered the theory of all aspects of the high resolution diffractometer, its aberrations and its adjustment in some detail, and given systematic methods for minimisation of the aberrations. We have not attempted to describe the current technology, for example of motor drives and angle encoders, since these will vary with manufacturer and with time. It is not necessary to have an understanding at the level of detail of this chapter for routine use of the instruments, but it is available as a resource, and should be found particularly useful when new experiments or new instruments are contemplated.

References

1. H. MARK, in Fifty years of X-ray diffraction, ed. P. P. EWALD (Int. Union of Crystallography, 1962) p. 603.
2. J. W. DUMOND, Phys. Rev., **52**, 872 (1937).
3. T. MATSUSHITA & U. KAMINAGA, J. Appl. Cryst., **13**, 465 and 472 (1980).
4. J. H. BEAUMONT & M. HART, J. Phys. E: Sci. Inst., **7**, 823 (1978).
5. K. KOHRA, M. ANDO, T. MATSUSHITA & H. HASHIZUME, Nucl. Inst. Meth., **152**, 161 (1978).
6. S. W. WILKINS & A. W. STEVENSON, Nucl. Inst. Meth., **A269**, 321 (1988).
7. R. W. JAMES, The optical principles of the diffraction of X-rays (Ox Bow Press, Connecticut, 1982) p. 308.

8. J. P. QUINTANA, M. HART, D. BILDERBACK, C. HENDERSON, D. RICHTER, T. SETTERSTON, J. WHITE, D. HAUSERMANN, M. KRUMREY & H. SCHULTESCHREPPING, J. Synchrotron Radiation, **2**, 1 (1995); J. P. QUINTANA & M. HART, J. Synchrotron Radiation, **2**, 119 (1995); L. E. BERMAN & M. HART, Nucl. Inst. Meth., **A334**, 617 (1993).

9. D. K. BOWEN, D. CHETWYND, N. KRYLOVA & S. SMITH, to be published.

10. N. LOXLEY, D. K. BOWEN & B. K. TANNER, Mat. Res. Soc. Symp. Proc., **208**, 107 (1991).

11. W. BARTELS, J. Vacuum Science and Technology, **B1**, 338 (1983).

12. N. LOXLEY, B. K. TANNER & D. K. BOWEN, Advances in X-ray Analysis, **38**, 361 (1995).

13. P. KIRKPATRICK, J. Opt. Soc. Am., **38**, 766 (1948); T. W. BARBEE, Opt. Eng., **25**, 898 (1986).

14. U. ARNDT, J. Appl. Cryst., **23**, 161 (1990).

15. M. SCHUSTER & H. GÖBEL, J. Phys. D: Appl. Phys., **28**, A270 (1995).

16. M. SCHWARZCHILD, Phys. Rev., **32**, 162 (1928).

17. N. LOXLEY, S. COCKERTON, B. K. TANNER, M. L. COOKE, T. GRAY & D. K. BOWEN, Mater. Res. Soc. Symp. Proc., **324**, 451 (1993).

18. P. FEWSTER, J. Appl. Cryst., **18**, 334 (1985).

19. B. K. TANNER, XI CHU & D. K. BOWEN, Mat. Res. Soc. Res. Symp. Proc., **69**, 191 (1986).

20. G. F. KNOLL, Radiation detection and measurement (Wiley, New York, 1979).

21. S. COCKERTON & B. K. TANNER, Adv. X-ray Anal., **38**, 371 (1995); S. COCKERTON, B. K. TANNER & G. DERBYSHIRE, Nuclear Instruments and Methods in Physics Research, **B97**, 561 (1995).

3

Analysis of Epitaxial Layers

In this chapter we discuss the measurement and analysis of simple epitaxial structures. After showing how to select the experimental conditions we show how to derive the basic layer parameters: the composition of ternaries, mismatch of quaternaries, misorientation, layer thickness, tilt, relaxation, indications of strain, curvature and stress, and area homogeneity. We then discuss the limitations of the simple interpretation.

3.1 Introduction

The greatest use of high resolution diffractometry in industry is without doubt the characterisation of epitaxial structures on compound semiconductors. Such epilayers are mismatched, misoriented, defective, non-uniform and bent. These defects affect device performance and production yield. Residual strains in the layer can be correlated with poor device performance or with degradation, and dislocations in the interface are particularly damaging. III–V layers are nowhere near as perfect as silicon, either in defect concentration or in composition uniformity. The defects affect carrier lifetimes and may act as non-radiative recombination centres, and the composition affects the bandgap as well as the mismatch. It is therefore essential in the development phase of a device to determine whether the defects generated in a particular production process will permit adequate yield of good quality devices. Later on, in manufacture, it is necessary to determine whether the process is under adequate quality control.

In this chapter we shall see how the basic parameters can be very easily obtained and interpreted from X-ray rocking curves for relatively simple structures. We define these as good quality substrates (giving rocking curves near the theoretical width) and up to, say, three layers, each at least $0.5\,\mu$m thick. Though these are simple, they nevertheless comprise a large sector of the optoelectronics device production and are certainly realistic cases. The discussion will also illustrate many important principles. In later chapters we shall develop the methods of interpretation of more complex cases.

3.2 Plane wave diffraction from heteroepilayers

Figure 3.1 shows schematically the defects that are common in epilayer structures. Mismatch is the most vital parameter to determine, both in itself as a measure of

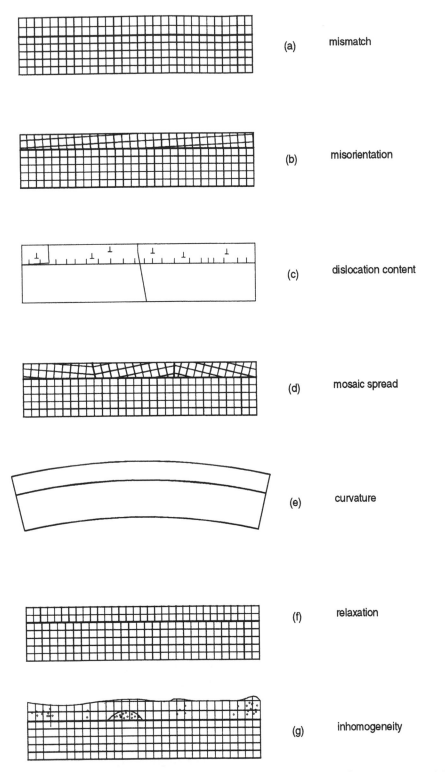

(a)   mismatch

(b)   misorientation

(c)   dislocation content

(d)   mosaic spread

(e)   curvature

(f)   relaxation

(g)   inhomogeneity

Figure 3.1 A schematic representation of the defects common in epilayer structures. (a) mismatch, (b) misorientation, (c) dislocation content, (d) mosaic spread, (e) curvature, (f) relaxation, (g) inhomogeneity

Table 3.1 The effect of substrate and epilayer parameters upon the rocking curve

Material parameter	Effect on rocking curve	Distinguishing features
Mismatch	Splitting of layer and substrate peak	Invariant with sample rotation
Misorientation	Splitting of layer and substrate peak	Changes sign with sample rotation
Dislocation content	Broadens peak	Broadening invariant with beam size
		No shift of peak with beam position on sample
Mosaic spread	Broadens peak	Broadening may increase with beam size, up to mosaic cell size
		No shift of peak with beam position on sample
Curvature	Broadens peak	Broadening increases linearly with beam size
		Peak shifts systematically with beam position on sample
Relaxation	Changes splitting	Different effect on symmetrical and asymmetrical reflections
Thickness	Affects intensity of peak	Integrated intensity increases with layer thickness, up to a limit
	Introduces interference fringes	Fringe period controlled by thickness
Inhomogeneity	Effects vary with position on sample	Individual characteristics may be mapped

strain in the epilayer but also because it allows us to determine composition for ternary compounds and hence infer the bandgap. Layer thickness is also an important characterisation parameter for device behaviour.

If we imagine the diffraction of a plane wave from epilayers we see that there will in general be differences of diffraction angle between the layer and the substrate, whether these are caused by tilt ($\delta\theta$) or mismatch (δd). Double or multiple peaks will therefore arise in the rocking curve. Peaks may be broadened by defects if these give additional rotations to the crystal lattice, and there will also be small peaks arising from interference between waves scattered from the interfaces, which will be controlled by the layer thicknesses. The material will show different defects in different regions. Table 3.1 summarises the influence on the rocking curve of the important parameters. The distinguishing features of each effect will be discussed quantitatively in this chapter. However, we need to ensure first of all that the experimental measurement is properly designed.

3.3 Selection of experimental conditions

Most high resolution diffractometry is done in reflection. III–V and II–VI materials are too absorbing to use conveniently in transmission and, in any case, reflection techniques provide information only from the relevant surface region.

In reflection, the intensity of the X-ray wavefield inside the crystal falls off very rapidly away from the surface, due to transfer of energy to the diffracted beam.

Absorption also becomes important at low incident angles to the surface. By choosing the radiation and the reflection (including its symmetry), the penetration may be varied between about 0.05 and 10 micrometres. This is ideally matched to device structures. This is quantified by the *extinction distance* ξ_g, defined as the depth at which the incident intensity has decreased to 1/e of its value at the surface. This may be calculated from diffraction theory, and some examples, for GaAs with CuK$_\alpha$ radiation, are shown in Table 3.2. It is assumed that the wafer surface is (001), hence the 004 reflection is symmetric and the others asymmetric.

The effect of absorption will be, roughly, to halve the extinction distances at the lower glancing angles. To get to really low extinction distances with characteristic radiation it is possible to use skew reflections to get very low glancing or grazing angles. The minimum penetration in diffraction is obtained at the critical angle for total external reflection. There is then a very rapid transition to external reflection; but, of course, there must be a little 'sampling' of the crystal by the beam and the penetration is a few atomic layers. This gives two-dimensional diffraction, which is now being used to give information about surface layers, but is rarely attempted with conventional X-ray sources. Even with more usual arrangements, glancing angle diffraction can be an invaluable tool for the study of thin films. Figure 3.2 shows the dramatic effect obtained for a thin epilayer of InGaAs on InP. We see that if the symmetrical reflection is not usable then the radiation and reflection should be chosen primarily to control penetration depth. Strong reflections give shorter extinction distances. Unfortunately, the strongest reflections of all in the zincblende structure, 111 and 220, are not accessible in reflection with (100) wafers with CuK$_\alpha$ radiation, though they are ideal in transmission.

Most of the important parameters may be derived from symmetric reflections, as shown below. The exception is the measurement of relaxation, that is, the extent to which the interface is less than perfectly coherent with the substrate. A large mismatch will cause a substantial strain in the layer which, at some critical thickness, will be relieved by the generation of interface dislocations. The average density of these may be accurately measured, by measuring the lattice mismatch parallel as well as perpendicular to the interface; this requires an asymmetric reflection with a substantial component of the reflection vector parallel to the interface, in addition to the symmetric reflection and the determination of layer tilt relative to the substrate.

The accuracy of all measurements will be improved if the background (away from the peaks) is low. This should never be more than a few cps if beams of 1×1 mm or less are used, and <0.1–0.2 cps is attainable with standard instruments and a good detector. The principles necessary to achieve this are as follows:

Table 3.2 Extinction distances in GaAs with CuK$_\alpha$ radiation

Reflection	Bragg angle (degrees)	Rocking curve width (arc seconds)	Extinction distance (μm)
004	33.024	5.34	7.1
044	50.420	6.80	9.5
115	45.070	2.60	21.2
224	41.873	9.50	6.5

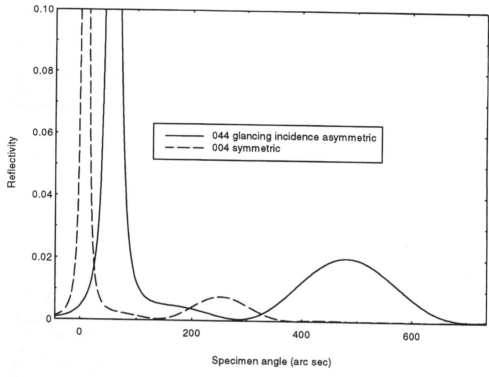

Figure 3.2 The effect of glancing incidence geometry, calculated for a 150 nm $In_{0.5177}Ga_{0.4823}As$ film on a (001) InP substrate. The symmetric 004 and the grazing incidence 044 curves are shown

1 Ensure that there is no unshielded path between the end of the collimator and the entrance of the detector.

2 Ensure that there is no unshielded path between the point at which the primary beam strikes the first crystal and the entrance of the detector.

A scatter shield must therefore surround the first crystal/collimator assembly, and the beam-defining slits should be placed in between the crystals. The necessity for defining the optic axis of the instrument, and placing the first and second crystal surfaces on this optic axis is now apparent. If high energy radiation is being used, it may be necessary to shield the sides of the detector as well. If problems remain, they can usually be overcome by experimenting with the placement of scatter shields using the above principles. Proper design of the point at which the collimator enters the tube shield is necessary as this can be a source of considerable noise. Lead-loaded Perspex (Plexiglass) is a very useful shielding material, since it is transparent and thus easy to align.

3.4 Derivation of layer parameters

We now assume that we have properly recorded rocking curves available, and that a substrate with a single epitaxial layer >0.5 μm thick (and less than, say, 5 μm) is measured. This will result in two peaks, one each from the substrate and layer. The same analysis will apply to multiple peaks (other than those from superlattices) provided that they are well separated so that interference effects are minimised. The basic parameters are derived as follows, with the symmetric reflection used unless otherwise specified. The examples are given for (001) substrates and layers, but are quite general (with the caution that if the Poisson ratio is required, the value appropriate to the crystal orientation should strictly be used).

3.4.1 *Mismatch*

The peak separation in the non-dispersive setting between the substrate and layer reflections is measured; call this $\delta\theta$. This is related to the change of interplanar spacing *normal* to the substrate through the equation

$$\delta d/d = -\delta\theta \cot\theta \qquad (3.1)$$

If the reflection is the usual symmetric 004 then the 'experimental X-ray mismatch'

$$m^* = \delta a/a = \delta d/d \qquad (3.2)$$

assuming, for the moment, that the layer is not tilted with respect to the substrate. This is derived directly from the measurements with no extra parameters or assumptions. However, this is not the mismatch that would be measured if the layer were removed from the substrate and allowed to relax to its natural, unstressed state. The epilayer is elastically constrained to match the substrate in the plane *parallel* to the substrate, in both x and y directions, and there is a consequent *tetragonal distortion* of the epilayer, as shown in Figure 3.3.

The true or relaxed mismatch m is defined with respect to the relaxed (i.e. the normally tabulated) lattice parameters a_l and a_s of the epilayer and substrate materials as

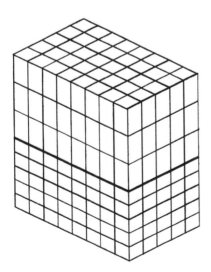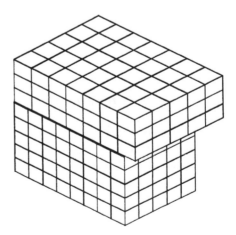

Figure 3.3 The tetragonal distortion in a coherent epilayer, (a) fully relaxed, and (b) constrained to match the substrate

$$m = \left(a_l - a_s\right)/a_s \tag{3.3}$$

It can then be calculated, by means of some straightforward elasticity theory[1] as

$$m = m * \left\{\frac{1-v}{1+v}\right\} \tag{3.4}$$

where m is the Poisson ratio; since $v \sim 1/3$, $m \sim m*/2$ as an approximate guide. For accurate work the Poisson ratios of the constituents must be determined, and in principle these will vary with the alloy content.

The intrinsic accuracy of mismatch measurements is ample. For example, a $1\,\mu m$ layer that is mismatched by 220 ppm from an InP substrate gives a peak splitting of 28″; this can be measured to about 2%, giving a resolution of 4–5 ppm. The peaks begin to merge below about 100 ppm but splitting down to about 50 ppm can be handled by simple methods; smaller splittings can be determined through computer simulation and fitting of the whole rocking curve as will be shown in Chapter 5.

3.4.2 *Composition*

For a ternary layer this may be determined from Vegard's law; this simply states that the lattice parameter of a solid solution alloy will be given by a linear dependence of lattice parameter on composition, following a line drawn between the values for the pure constituents. This process may be automated through software that performs all calculations necessary to extract peak splitting, relaxed mismatch and composition automatically for ternary layers. Vegard's law[2] was originally proposed for ionic salt pairs, e.g. KCl–KBr but has been widely investigated for metals, in which it does not work too well, and widely assumed for III–V semiconductors. It is based on elastic interactions between atoms, and is thus reasonable when all electronic interactions in the alloy series are very similar.

Quaternary multilayers present a substantial challenge because there are so many variables. If the layers are not graded, and if there are only two types, e.g. InP substrate and buffer layers and a single quaternary composition for the active layer and cap, the mismatch may be obtained from the peak splitting. The composition cannot be obtained from the mismatch since a degree of freedom remains. However, a photoluminescent measurement of the bandgap plus the mismatch measurement does suffice.

3.4.3 *Substrate misorientation*

Substrates are often specified at some controlled angle from (001), and this may need to be verified. A specimen stage is required that can rotate the specimen about its surface normal. If the specimen rotation method is used to bring the reflecting planes of the specimen parallel to those of the reference crystal then in the final configuration the reflecting plane normals in the specimen and reference crystals and the specimen surface normal are all coplanar. Rotation of the specimen about a further 180° (which is the other coplanar setting) will give a shift in the position of the Bragg peak. This will be exactly twice the misorientation angle between the reflecting plane and the specimen surface.

56

3.4.4 Tilt

If the layer is tilted relative to the substrate then this will result in a shift of the layer peak relative to that of the substrate for reasons unconnected with composition. We have two problems – how to measure true splittings, and how to determine the tilt itself. The peak shift due to tilt will vary with the absolute direction of the incident beam relative to the substrate (i.e. with respect to rotation of the specimen about its surface normal), and may thereby be distinguished from mismatch splitting. If the specimen is rotated α about its normal, the displacement of the layer peak from the position it would have were there no tilt is $\beta \cos \alpha$, where β is the angle of tilt. It follows that a true splitting mismatch may be taken by rotating the specimen 180° in its plane and averaging the two measurements of splitting, i.e.

$$\delta\theta = \left(\delta\theta_0 + \delta\theta_{180} \right)/2 \tag{3.5}$$

This is insufficient to measure the tilt itself since we do not know whether our original measurement was in the direction in which the tilt would be a maximum (i.e. with the reflecting plane normals in the reference crystal, substrate *and layer* all coplanar). A third measurement (at least) is required, followed by fitting the three measurements of layer peak deviation to a sine curve to find the maximum deviation – this is then the tilt value. The best strategy is to make the third measurement at 90° to the first. We then have measurements at values of α displaced by 0°, 90° and 180°. We do not yet know where the origin of α is; let this be at $-\omega$ from our zero value of α. Then after extracting the mismatch part of splitting as shown above, we have three values of splitting, Δ_0, Δ_{90} and Δ_{180} which are given in terms of β (the tilt) and ω (the zero position of the tilt, when all diffracting vectors are coplanar) as follows:

$$\Delta_0 = \beta \cos \omega \tag{3.6}$$

$$\Delta_{90} = \beta \cos(90 + \omega) = -\beta \sin \omega \tag{3.7}$$

$$\Delta_{180} = \beta \cos(180 + \omega) = -\beta \cos \omega \tag{3.8}$$

then from (6) and (7)

$$\tan \omega = \Delta_{90}/\Delta_0 \tag{3.9}$$

which determines ω. The tilt β can then be found by substitution in any of the equations.

3.4.5 Dislocation content

Dislocations are commonly present in two regions. A layer with high mismatch may relax so that interface dislocations are created to accommodate the strain. A network at the interface is thus observed. Slip dislocations may be generated by local plastic deformation due to thermal or mechanical strain and propagate elsewhere in the layer. Dislocations in the layer itself may also be generated during the growth process, due, for example, to the presence of inclusions.

Interface dislocations give a specified relaxation of strain between the substrate and the epilayer, which gives quantifiable shifts in the positions of peaks in asymmetric reflections as discussed later in this chapter. As structures become more complex it is difficult to know which effects may be ascribed to interface relaxation and which to the layer structure itself. It is therefore often very useful to perform topography to see the dislocations directly, as discussed in Chapters 8–10.

On the other hand, dislocations inside the epilayer may have any of the possible Burgers vectors, and will on average contain roughly equal numbers of each sign. These do not shift the rocking curve, but they both broaden it and add diffuse scattering. A simple model for the broadening was given by Hirsch,[3] who showed that a reasonable estimate for the dislocation density ρ is

$$\rho = \frac{\beta^2}{9b^2}$$

(3.10)

in cm^{-2}, where β is the broadening of the rocking curve in radians and b is the Burgers vector in centimetres. Kaganer *et al.*[4] have recently published a much better quantitative model for the broadening.

The diffuse scatter arises because dislocations are defects which rotate the lattice locally in either direction. This gives rise to scatter, from near-core regions, which is not travelling in quite the same direction as the diffraction from the bulk of the crystal. This adds kinematically (i.e. in intensity not amplitude) and gives a broad, shallow peak that must be centred on the Bragg peak of the dislocated layer or substrate since all the local rotations are centred on the lattice itself. We can model the diffuse scatter quite well by a Gaussian or a Lorentzian function of the form:

$$I = \frac{A}{k^2 + \Delta\theta^2}$$

(3.11)

A further effect that can be seen in thick layers with large mismatch, for example GaAs on Si, is the decrease of dislocation density from interface to surface; this is aimed for by the crystal growers, and gives a rocking curve that is hard to simulate. However, an indication of the decrease in dislocation density can be obtained using two measurements, one a usual symmmetric 004 and the other highly asymmetric to confine the beam to a region much nearer the surface.[5] If the dislocation content indeed decreases towards the surface, then rocking curves taken with smaller extinction distances (and the same area measured) will show less broadening.

3.4.6 *Curvature and mosaic spread*

If the specimen is curved, then the change in angle from one side of the beam to the other will be cross-correlated with the rocking curve, and the latter will be broadened. An estimate of this effect is straightforward. Let the radius of curvature of the specimen be R and the diameter of the beam be s. The angular change of the incident angle across the beam will then be s/R, as illustrated in Figure 3.4. For standard measurements, we want this to be a small fraction of the intrinsic rocking curve width; with III–V materials we might accept a broadening of 2″ or 10^{-5} radians, which is 10–15% of the intrinsic width with CuK$_\alpha$ radiation. If the beam diameter is 1 mm then our criterion is satisfied if the specimen radius of curvature is no less than 100 m (Table 3.3). This is not always true and emphasises the need for small beams if accurate double-axis rocking curve widths are to be obtained on stressed specimens, though the splitting is not affected.

It is possible to compensate for curvature in theoretical simulation of the rocking curve. If the beam profile is square, then the rocking curve is simply correlated with

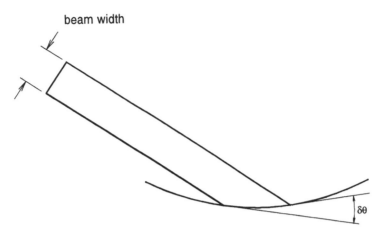

beam width

δθ

Figure 3.4 The change of incident angle across the beam on a curved specimen

Table 3.3 The effect of curvature on rocking curve broadening

Beam size (mm)	Radius of curvature (m)	Broadening (arc seconds)
5	100	10
5	20	50
1	100	2
1	20	10
0.5	100	1
0.5	20	5

a box function whose width is the angular change across the beam, and this is usually a good enough approximation. However, it is a consequence of the convolution process that some information is lost – for example, fine interference fringes will be washed out. The curvature for a given mismatch is less serious on small specimens than on complete wafers, so if stress is suspected and details are required, the measurement can be repeated on a cleaved-out portion of the specimen.

The curvature itself can be quite easily measured by translating the specimen a distance x in its plane along a diameter (a circular wafer is assumed), repeating the measurement and noting the shift $\delta\theta$ in the absolute position of the Bragg peak. Then, again,

$$R = s/\delta\theta \qquad (3.12)$$

This measures the curvature about an axis perpendicular to the dispersion plane, i.e. the cylindrical curvature, and it may be necessary to rotate the wafer through 90° to get the orthogonal component. This may be related to absolute stress in the wafer with knowledge of the wafer thickness, diameter and elastic modulus. The most accurate method is to measure a number of points on a wafer and use a linear regression formula for the average curvature.

If the specimen has been rotated (Figure 3.5) then we have

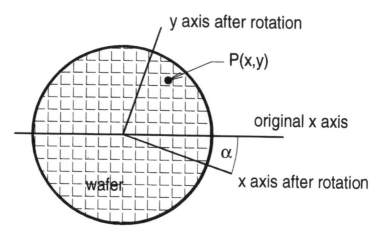

Figure 3.5 The projection of measured points onto the dispersion plane

$$s = x \cos \alpha + y \sin \alpha \qquad (3.13)$$

$$\theta_{substrate} = \left[\frac{1}{R}\right] s + \theta_0 \qquad (3.14)$$

The linear regression (least squares) is performed on $\theta_{substrate}$ and s to get $1/R$. The residuals give the uniformity of curvature.

3.4.7 *Relaxation*

Equation (3.4) contains the assumption that the interface is fully coherent. If it is only partially coherent, i.e. it contains interface dislocations, it is said to be relaxed, and equation (3.4) is not valid for the determination of the relaxed mismatch. Note the two different usages of the word 'relaxed' in the last sentence! It is necessary to measure the misfit parallel to the interface as well as perpendicular. For this, we need an asymmetric reflection which is at as high an angle to the surface as possible: 224 and 113 are both acceptable.

Figure 3.6 shows a coherent and a relaxed layer, and it is clear that both the mismatch and the misorientation change between the substrate and the layer. The *tetragonal distortion* changes. From equations (3.6) and (3.9) it is clear that the effect of tilt on the splitting is reversed if the specimen is rotated by 180° about its surface normal, but the splitting due to the mismatch will not be affected by such a rotation. Thus we may make grazing incidence or grazing exit measurements, Figure 3.7, to separate the tilt from the true splitting.

The resulting measured splittings $\Delta\theta_i$ and $\Delta\theta_e$ are now different between these two geometries:

Grazing incidence:

$$\Delta\theta_i = \delta\theta + \delta\phi \qquad (3.15)$$

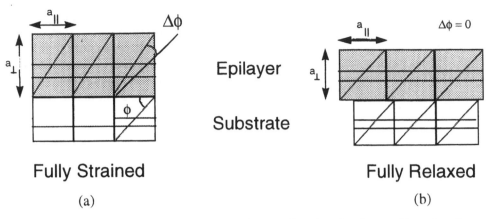

Fully Strained

(a)

Fully Relaxed

(b)

Figure 3.6 A side view of (a) coherent and (b) partially relaxed epilayers. The relaxation process changes both the interplanar spacings of the epilayer and the angles between the reflecting planes and the surface. (Courtesy K. M. Matney)

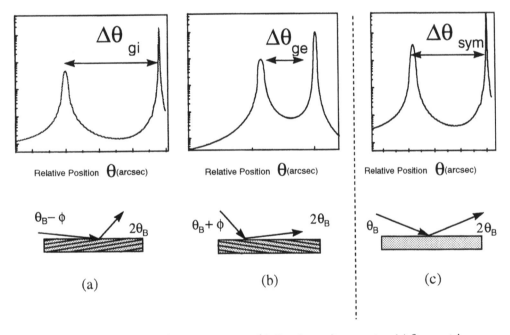

Figure 3.7 (a) Grazing incidence geometry. (b) Grazing exit geometry. (c) Symmetric scan. (Coustesy K. M. Matney)

Grazing exit:

$$\Delta\theta_i = \delta\theta - \delta\phi \tag{3.16}$$

Thus we may determine $\delta\theta$ and $\delta\phi$ independently. We need to know the lattice parameter of the layer parallel and perpendicular to the substrate, i.e. the *in situ* cell parameters a_l, b_l and c_l of the layer. From these we may calculate the relaxation and (together with the Poisson ratio) the fully relaxed lattice parameter of the epilayer. Let the lattice parameter of the (cubic) substrate be a_s, let hkl be the (asymmetric) reflection, λ the wavelength, θ the Bragg angle and ϕ the angle between the reflecting plane and the surface, with subscripts s and l referring again to the substrate and epilayer. We consider first the case in which the relaxation is symmetrical, i.e. the distortion is tetragonal and $a_l = b_l$. We know λ, θ_s and ϕ_s. Using equations (3.10) and (3.11) we may therefore calculate

$$\theta_l = \theta_s + \delta\theta \tag{3.17}$$

$$\phi_l = \phi_s + \delta\phi \tag{3.18}$$

From the equation for the interplanar spacing in tetragonal crystals,

$$\frac{1}{d_{hkl}^2} = \frac{h^2 + k^2}{a_l^2} + \frac{l^2}{c_l^2} \tag{3.19}$$

and the Bragg law, we obtain

$$\frac{4\sin^2\theta_l}{\lambda^2} = \frac{h^2 + k^2}{a_l^2} + \frac{l^2}{c_l^2} \tag{3.20}$$

Assuming a 001 crystal surface, we use the formula for interplanar angles in tetragonal crystals to obtain

$$\sec^2\phi_l = \frac{c_l^2}{l^2}\left\{\frac{h^2 + k^2}{a_l^2} + \frac{l^2}{c_l^2}\right\} \tag{3.21}$$

From equations (3.16) and (3.17) we may solve for the cell constants of the layer:

$$c_l = \frac{l\lambda}{2\sin\theta_l\cos\phi_l} \tag{3.22}$$

and

$$a_l = \frac{l\lambda}{2\sin\theta_l}\sqrt{\frac{h^2 + k^2}{l^2}} \tag{3.23}$$

The accuracy of c_l may be checked from measurement of the symmetric 001 reflection. The relaxation R (%) is defined as

$$R = \frac{a_l - a_s}{a_l^R - a_s} \times 100 \tag{3.24}$$

where a_l^R is the fully relaxed lattice parameter (cubic) of the epilayer. We may find this if we know the Poisson ratio of the layer, v, since for any layer strained in two coplanar directions ε_{xx} and ε_{yy} the vertical strain is given by[1]

$$\varepsilon_{zz} = -\left(\varepsilon_{xx} + \varepsilon_{yy}\right)\left\{\frac{v}{1-v}\right\} \tag{3.25}$$

Since $\varepsilon_{xx} = \varepsilon_{yy} = (a_l^R - a_l) / a_l^R$, and $c_l = a_l^R(1 + \varepsilon_{zz})$,

$$c_l = a_l^R\left[1 - \frac{2\left(a_l^R - a_l\right)}{a_l^R}\left\{\frac{v}{1-v}\right\}\right] \tag{3.26}$$

$$a_l^R = \frac{c_l - 2a_l\left(\dfrac{v}{1-v}\right)}{1 - 2\left(\dfrac{v}{1-v}\right)}$$

(3.27)

The fully relaxed lattice parameter of the layer, a_l^R, is the value that we need to use in Vegard's law to find the composition of the epilayer.

If we are confident that the (001) planes of the epilayer and substrate remain parallel, as is reasonable if the substrate is cut parallel to (001) – but not if it is offcut – we can in principle derive the relaxation from just the symmetric and the grazing incidence reflections, without using the grazing exit. This may be important in quality control to save measurement time. The derivation is lengthy but straightforward. The symmetric reflection provides c_l. We find expressions for $\delta\theta$ and $\delta\phi$ in terms of the lattice parameters of the substrate and of and c_l, using the tetragonal interplanar spacing and angle formulae. We solve this for a_l, which gives

$$a_l = \frac{\sqrt{\left(h^2 + k^2\right)c^2}}{\sqrt{\left\{\dfrac{l\,\mathrm{cosec}\,\phi - \dfrac{c\tan\theta\sqrt{h^2 + k^2 + l^2}}{a_s}}{\Delta\theta - \tan\theta + \cot\phi}\right\} - l^2}}$$

(3.28)

and we then proceed as before. This equation, which has not previously been published, contains the approximations that $\delta\theta$ and $\delta\phi$ are small, and if this is not so an iterative solution is preferable.[5] Beginning with the known c_l and the value for a_l given by equation (3.27), we calculate $\Delta\theta$ and iterate a_l until the agreement with the experimental $\Delta\theta$ is good.

Once the parallel mismatch is determined, some information about average dislocation density in the interface may be obtained. It is not possible unambiguously to determine the types of dislocation present, since different types of dislocation may combine to give the same strain. However, the parallel mismatch is entirely due to dislocation content in (or very near) the interface and thus

$$\frac{a_l - a_s}{a_s} = \frac{b}{s}$$

(3.29)

where b is the magnitude of the Burgers vector of the interface dislocations projected in the direction given by the intersection of the incidence plane with the interface, and s is their spacing in the same direction. If the nature of the dislocations is known, for example from electron microscopy or X-ray topography, and two asymmetric reflections are taken, the overall interface density can be determined unambiguously.

3.4.8 *Thickness*

The layer thickness determines the relative intensity of the layer and substrate peaks. Provided again that the structure is simple (later chapters deal with cases that are not), then the intensity of the layer peak increases monotonically with thickness. The values and the calibration constant will depend on the particular system being measured and can be determined either empirically or by computer calculation. An example is shown in Figure 3.8.

The proper value to use for intensity measurements is the integrated intensity – the area under the peak – rather than the peak intensity. This is because it is less variable with material structure. If many dislocations are present, for example in a strained layer material such as InGaAs on GaAs of moderate thickness, then the layer peak will be lowered and broadened, but to first order the integrated intensity will be the same. It is not always possible to measure the integrated intensity when peaks are close together. The measurement of integrated intensity is taken out to the tails of the peak where it is indistinguishable from background. Clearly, better values will be obtained if the background is low, which emphasises the need for good experimental technique.

The rocking curves of Figure 3.8 show small periodic oscillations around the layer peak. These are called thickness fringes and are a very good way of measuring layer thickness. The most accurate way is through computer simulation (Chapter 5), but it is often useful to get a start to the simulation by measuring the interference peak separation, $\Delta\theta_p$, which is given by

$$\Delta\theta_p = \frac{\lambda\gamma_g}{t\sin 2\theta}$$

(3.30)

where λ is the wavelength, t the thickness and γ_g the cosine of the angle between the diffracted beam and the inward-going surface normal. For the reflection case, this may be expressed as

$$\Delta\theta_p = \frac{\lambda\sin(\theta\pm\phi)}{t\sin 2\theta}$$

(3.31)

where ϕ is again the angle between the reflecting plane and the surface, the positive sign applies to grazing incidence and the negative sign to grazing exit. For the common symmetrical case, we may simplify and rearrange to obtain

$$t = \frac{\lambda}{2\Delta\theta_p\cos\theta}$$

(3.32)

This very useful method also has the advantage that the equations do not contain anything about the material or diffraction conditions other than the Bragg angle and geometry. The independence from material parameters arises because the refractive index for X-rays is very close to unity. The equations are, of course, similar to those for optical interference from thin films, since the physics is the same, but in the optical case we do need to know the refractive index.

If more than one layer is present, there will be interference fringes from each of them. If their thicknesses are different these will superimpose and beat. Very complex patterns may arise, as will be seen in the HEMT structure simulated in Chapter 5.

Newcomers to the field often assume that a simple Fourier analysis will suffice to pull out the periodicities automatically. However, the problem is not at all simple, as has been shown by Hudson, Tanner and Blunt.[7] There are two problems. One is that the interference fringes are quite weak, and relatively few are observed. The

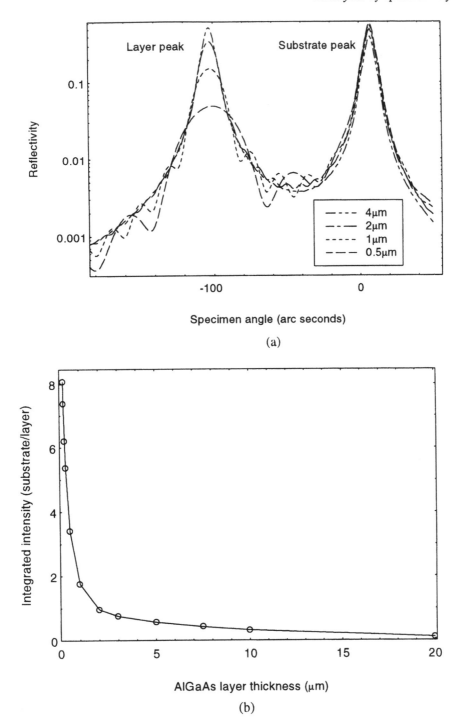

Figure 3.8 The effect of thickness on layer peak intensity. (a) Rocking curves, (b) integrated intensity ratio between substrate and epilayer peaks. Calculated curves for $Al_{0.3}Ga_{0.7}As$ on GaAs. CuK_α radiation 004 reflection

Table 3.4 Comparison of best Fourier analysis with specimen structure[7]

Layer thickness from matching to simulation	Thickness extracted from Fourier transform
550 Å GaAs	990 ± 20 Å
400 Å $Al_{0.18}Ga_{0.82}As$	1100 ± 20 Å
180 Å $In_{0.13}Ga_{0.87}As$	

intensity of the Fourier peak is therefore low, and may be obscured by noise. A low-noise detector is essential. The second is that the transform is absolutely dominated by the sharp substrate peak, which of course puts intensity throughout Fourier space and masks any other modulation. The procedure that showed most success was

1 Mask out the substrate peak from the rocking curve; in practice, truncate the curve and study only one side.

2 From the average values of the curve determine a 'background' to the thickness oscillations, and divide the data by the smoothed curve.

3 Subtract the residual background.

4 Perform an autocorrelation on the processed curve.

5 Perform the Fourier transform.

The results of the analysis are shown in Table 3.4. It is seen that, although some of the periodicities are extracted, some which appear are difference or beat frequencies and do not correspond to a single layer. The conclusion is that Fourier analysis, with the above procedure, is a powerful aid to a skilled researcher, but it is not yet appropriate for automated analysis.

More complex integral transforms have been studied, e.g. the two-dimensional Wigner transform,[8] but so far without success. Even the best-quality data are too sparse and noisy to give reliable information when transformed in this way.

3.4.9 Area homogeneity

Whatever the crystal growers claim, epitaxial layers are not uniform across their area! One per cent consistency is very good. It is therefore necessary to repeat each measurement at a number of points across each wafer. A 9×9 grid across a 3″ wafer is thorough, whilst a 3×3 grid often suffices. Clearly, automated data collection over such a grid is a great help. Figure 3.9 illustrates a 'wafer map' plotted as a mesh surface. This may be applied to any parameter that can be extracted quantitatively from the rocking curve.

3.4.10 Limitations and problems

In the whole of this chapter we have taken an 'intensity' viewpoint, which requires that frequencies are well separated and interaction is minimal. We have shown that a very large amount of information may be derived in this way. But of course, the

Composition

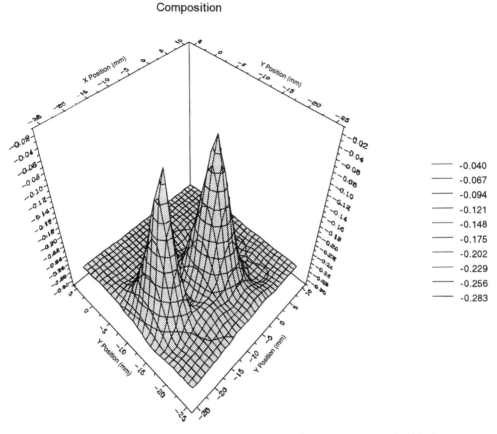

Figure 3.9 Surface mesh plot showing the variation of In content in an InAlAs layer on GaAs. (Courtesy T. Lafford)

true physics of waves must consider amplitude and not intensity, and we should expect that interference effects dominate in certain situations. These are in cases of

- thin layers, say below $0.5\,\mu$m
- small mismatches, so that peaks overlap
- multiple peaks
- superlattices

Computer simulation is required to analyse such cases, as will be discussed in later chapters.

3.5 Summary

In this chapter we have seen what information can be derived directly from the rocking curve, without computer simulation. This has laid the foundation for more detailed interpretation. This approach is also of immediate practical application in the semiconductor industry, since the procedures discussed may mostly be auto-mated in software and used for very rapid analysis in quality control applications. It is perfectly possible to measure and to analyse the composition of a ternary epilayer, automatically, in less than thirty seconds.

References

1. L. D. LANDAU & E. M. LIFSCHITZ, Elasticity (Pergamon, Oxford, 1972).
2. L. VEGARD, Z. Physik, **5**, 17 (1921).
3. P. B. HIRSCH, 'Mosaic structure', Chapter 6 in Progress in metal physics, eds. B. CHALMERS & R. KING (Pergamon, New York, 1956).
4. V.M. KAGANER, R. KÖHLER, M. SCHMIDBAUER, R. OPITZ & B. JENISCHEN, Phys. Rev B., **55**, 1973 (1997).
5. J. W. LEE, D. K. BOWEN & J. P. SALERNO, Mat. Res. Soc. Symp. Proc., **91**, 193–8 (1987).
6. K. M. MATNEY, Mat. Res. Soc. Symp. Proc., **379**, 257 (1995).
7. J. M. HUDSON, B. K. TANNER & R. BLUNT, Adv. X-ray Analysis, **37**, 135 (1994).
8. R. CLINCIU, MSc thesis, University of Warwick (1992).

4

X-ray Scattering Theory

In this chapter we first extend the kinematical theory, discussing the strengths of X-ray reflections, forbidden reflections, the thin crystal solution, anomalous dispersion and reciprocal space geometry. We then review the results of the dynamical theory: deviation parameters, dispersion space and geometry, boundary conditions, the range of Bragg reflections, polarisation, intensity formulae and penetration depth. We briefly discuss spherical wave theory, and Penning–Polder theory for distorted crystals. Finally we give a summary of useful formulae.

4.1 Introduction

The theory of X-ray scattering is highly practical! It is an accurate theory, based on a few sound assumptions, and with implementations on personal computers it may be used to interpret the structures of advanced industrial materials and thereby to assist in process development and quality control. The characterisation scientist or engineer who has a good grasp of the theory will therefore be able to design better measurements and to interpret them more accurately.

It is not our intention to provide full derivations of X-ray scattering theory, since this is primarily of interest to the specialist and may be found in many excellent books and reviews. The characterisation scientist needs a qualitative understanding in order to appreciate general features of experiment design and the interpretation of high resolution rocking curves and images. He or she also needs specific numbers such as the ideal rocking curve width or the penetration depth for a particular specimen. Our aim is therefore simply to explain the aspects of the theory that are relevant to later chapters and to summarise the important formulae.

It is conventional and useful to approach X-ray scattering theory on two levels, the so-called *kinematical* and *dynamical* theories. The simpler kinematical theory assumes that a negligible amount of energy is transferred to the diffracted beam, with the consequence that we can ignore rediffraction effects. This is fairly accurate for the geometry of diffraction in all cases, and, reasonably enough, is also fairly accurate for the intensities when the scattering is very weak. Very thin crystals, surface scattering and diffuse scattering are examples of weak scattering. When the scattering is strong, this assumption is hopeless, for example for the diffracted intensities and rocking curve widths of near-perfect crystals. Examples of these are the majority of the semiconductor structures used in industry so it is easy to see why we must use the dynamical theory for characterisation research. Fortunately, many of the concepts of kinematical theory, such as structure factor and diffraction geometry, are also used in the dynamical theory.

4.2 Kinematical theory

4.2.1 *The structure factor*

We began the discussion of kinematical theory in Chapter 1, showing how the scattering from atoms is added up with regard to the phase to form the scattering from a unit cell, the structure factor. We repeat this important equation here. The structure factor for the *hkl* reflection is

$$F_{hkl} = \sum_i f_i \exp\left\{-2\pi i\left(hu + kv + lw\right)\right\}$$

(4.1)

where (*uvw*) are the fractional coordinates of the vector **r**, which runs from the origin of the unit cell to the atom of type *i* whose atomic scattering factor is f_i and the summation is over all atoms in the unit cell. Let us now look more deeply at this equation and consider the assumptions made. These are:

1   The scattered intensity is very small. The loss of intensity due to re-scattering is negligible and thus the refractive index is unity.

2   The point of observation is at a large distance compared with the dimensions of any coherently illuminated scattering volume.

3   Scattered waves from different atoms are nearly parallel. We label these with the single wavevector $\mathbf{k}_h$.

Conditions (2) and (3) are equivalent to the Fraunhofer or far-field approximations in ordinary optics. The coherently illuminated region with usual laboratory X-ray sources is a few micrometres across. We therefore expect this theory to be useful in the cases of weak scattering but to be seriously awry for strong scattering.

The atomic scattering factors f_i are usually calculated in terms of the scattering of an individual free electron. This is calculated as if the electron were a classical oscillator – since the assumption is that the electron is a free charged particle. It is set into forced oscillation by the radiation field of an incident X-ray and then re-radiates in all directions at the same frequency as the incident wave frequency. This is termed elastic or Thompson scattering. For an X-ray beam, the intensity *I* scattered by one electron through an angle 2θ relative to the incident intensity I_0 is

$$\frac{I}{I_0} = \frac{C^2 r_e^2}{R^2}$$

(4.2)

where *R* is the distance of observation from the particle and r_e (the so-called classical electron radius) is (in cgs units, in which this theory is most concisely expressed)

$$r_e = \frac{e}{mc^2}$$

(4.3)

where e is the electronic charge, m the rest mass of the electron and c the velocity of light. C is a factor dependent on the polarisation. If the electric vector of the X-ray wave is perpendicular to the dispersion plane then $C = 1$; this is known as σ polarisation. For π polarisation, in which the electric vector is parallel to the dispersion plane, $C = \cos 2\theta$. For an individual electron, the angular variation of the scattering only arises through this polarisation term.

It does not matter that electrons in a solid are not classical oscillators, since this is only a unit of reference, but it brings out the polarisation behaviour. In fact, electrons in an atom behave surprisingly similarly to classical oscillators with respect to elastic X-ray scattering. If all the electrons in an atom were concentrated at one point, then we should just multiply the electron scattering factor f by Z, the atomic number, to get the atomic scattering factor f_i. This is a good approximation when the scattering angle 2θ is small and thus all the scattering is nearly in phase. However, atoms have finite sizes compared with X-ray wavelengths and when $2\theta \neq 0$ we must add with regard to phase, just as we added the scattering of atoms with regard to phase to arrive at the structure factor, equation (4.1). We shall have a similar equation for the atomic scattering factor, but since the distribution of electrons is continuous we express it as an integral rather than a summation. In units of f, or *electron units* (and hence itself dimensionless), this is:

$$f_i = \int_{space} \rho(r) \exp(2\pi i \mathbf{Q.r}) \, dV \tag{4.4}$$

The scattering vector $\mathbf{Q} = \mathbf{k}_h - \mathbf{k}_0$ where $\mathbf{k}_0$ is the incident beam vector and $\mathbf{k}_h$ the scattered beam vector, as illustrated in Figure 4.1.

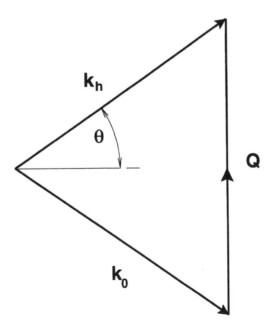

Figure 4.1 The vector relationship between the incident beam vector $\mathbf{k}_0$, the scattered beam vector $\mathbf{k}_h$ and the scattering vector $\mathbf{Q}$. The directions of the vectors correspond to their real directions, and the magnitudes of $\mathbf{k}_0$ and $\mathbf{k}_h$ are both $1/\lambda$

It is seen from Figure 4.1 that the modulus of the scattering vector, Q, is given by

$$Q = \frac{2\sin\theta}{\lambda} \tag{4.5}$$

In equation (4.4), $\rho(r)\,dV$ is the probability that an electron lies in a volume element dV of the atom at a radial distance r from the nucleus. $\rho(r)$ is the electron probability density and is a meaningful and measurable quantity. The scattered amplitude is therefore the Fourier transform of the electron density. We see immediately that for a given atom the atomic scattering factor is a function only of $\mathbf{Q}$. Its angular dependence is important and is shown in Figure 4.2.

4.2.2 *The strengths of X-ray reflections*

The best equations to use for intensity are those of the dynamical theory, but kinematical theory gives some useful insights, based simply upon the idea that the scattering increases monotonically with the structure factor, equation (4.1).

The structure factors of reflections in different cubic structures are shown in Table 4.1. Note that the structure factors may be complex, but this depends upon the choice of origin of the unit cell. The intensity formulae always contain the modulus

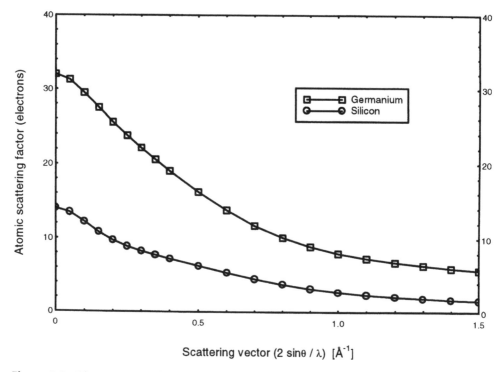

Scattering vector (2 sinθ / λ) [Å^{-1}]

Figure 4.2 The variation of the atomic scattering factor f_i with scattering angle 2θ. Values for silicon ($Z = 14$) and germanium ($Z = 32$) are shown

Table 4.1 Values of $|F_{hkl}|$ for a number of useful reflections in cubic structures, with examples of $|F_{hkl}|/V$ in electron units per $Å^3$ for one crystal of each type

Reflection	fcc	Al	Diamond cubic	Si	Sphalerite	GaAs
001	0	0	0	0	0	0
002	$4f$	0.41	0	0	$4(f_1 - f_2)$	0.05
004	$4f$	0.28	$8f$	0.39	$4(f_1 + f_2)$	1.02
111	$4f$	0.43	$5.66f$	0.38	$4\sqrt{f_1^2 + f_2^2}$	0.99
222	$4f$	0.31	0	0	$4(f_1 - f_2)$	0.04
333	$4f$	0.21	$5.66f$	0.24	$4\sqrt{f_1^2 + f_2^2}$	0.61
011	0	0	0	0	0	0
022	$4f$	0.35	$8f$	0.45	$4(f_1 + f_2)$	1.20
044	$4f$	n.a.	$8f$	0.31	$4(f_1 + f_2)$	0.80
112	0	0	0	0	0	0
224	$4f$	0.23	$8f$	0.35	$4(f_1 + f_2)$	0.90
113	$4f$	0.32	$5.66f$	0.30	$4\sqrt{f_1^2 + f_2^2}$	0.79
115	$4f$	0.21	$5.66f$	0.24	$4\sqrt{f_1^2 + f_2^2}$	0.61

of the structure factor $|F_h| = \sqrt{F_h F_h^*} = \sqrt{F_h F_{\bar{h}}}$ (or its equivalent in susceptibility, e.g. $\sqrt{\chi_h \chi_{\bar{h}}}$), and also $1/V$, where V is the volume of the unit cell. The modulus formulae are given in this table. Also given are the actual moduli for aluminium, silicon and gallium arsenide, to show the effects of material and of the scattering angle; these have been divided by V to make direct comparison between the materials. A number of the values are zero, for example the 001 and 002 reflections in the diamond cubic structure possessed by silicon and germanium. These are often referred to as forbidden reflections. Reflections such as 002 and 222 in sphalerite, the structure of a very important class of semiconductors such as gallium arsenide and indium phosphide, are not forbidden but are weak, since they depend upon the difference between the atomic scattering factors of the constituent atoms. They are sometimes called 'quasi-forbidden', and are very useful for emphasising compositional differences in such materials, such as non-stoichiometry or superlattice structures. An example is the common GaAs/GaAlAs superlattice, in which the 002 reflection is most useful.

4.2.3 *Intensity diffracted from a thin crystal*

This is the most useful quantitative intensity formula that may be derived from kinematical theory, since it is applicable to thin layers and mosaic blocks. We add up the scattering from each unit cell in the same way that we added up the scattering from each atom to obtain the structure factor, or the scattering power of the unit cell. That is, we make allowance for the phase difference $\mathbf{r}_i \cdot \mathbf{Q}$ between waves scattered from unit cells located at different vectors $\mathbf{r}_i$ from the origin. Quantitatively, this results in an interference function J, describing the interference of waves scattered from all the unit cells in the crystal, where

$$J = \sum_i \exp\left(2\pi i\, \mathbf{r}_i \cdot \mathbf{Q}\right) \tag{4.6}$$

The summation is taken over all unit cells in the crystal. The overall scattering amplitude J is given by $A = F_{hkl}J$ (where F_{hkl} is the structure factor for the hkl reflection) and the intensity I by the square of the amplitude

$$I = F_{hkl}^2 J^2 \tag{4.7}$$

We may evaluate this for a small crystal. In order to consider the size effects, we describe the crystal as a parallelepiped of side $n_1\mathbf{a}_1$, $n_2\mathbf{a}_2$ and $n_3\mathbf{a}_3$, where $\mathbf{a}_1$, $\mathbf{a}_2$ and $\mathbf{a}_3$ are the unit vectors defining the unit cell and the n_i are the number of unit cells in each side of the parallelepiped. The interference function becomes

$$J = \sum_{n_1=1}^{n_1} \exp\left(2\pi i\, n_1\mathbf{a}_1 . \mathbf{Q}\right) \sum_{n_2=1}^{n_2} \exp\left(2\pi i\, n_2\mathbf{a}_2 . \mathbf{Q}\right) \sum_{n_3=1}^{n_3} \exp\left(2\pi i\, n_3\mathbf{a}_3 . \mathbf{Q}\right) \tag{4.8}$$

We know from the simple description in Chapter 1 that we shall not get significant intensity unless the Bragg law is satisfied, so we shall formulate the problem so that we look at deviations from the exact Bragg condition. From equation (4.5), the latter is given when $Q = 1/d_{hkl}$, where d_{hkl} is the spacing of a plane with Miller indices hkl, and whose structure factor is not zero. In the notation of Figure 4.1, the Bragg geometry and Bragg law are satisfied if the vector $\mathbf{Q} = \mathbf{h}$, where $|\mathbf{h}| = 1/d_{hkl}$, and the direction of $\mathbf{h}$ is perpendicular to the (hkl) planes (we shall see shortly that this is saying that $\mathbf{h}$ is a vector of the *reciprocal lattice* of the crystal). Simple trigonometry on Figure 4.1 then gives the Bragg law. We can express deviations from the Bragg law in terms of a deviation vector $\mathbf{q}$, i.e.

$$\mathbf{Q} = \mathbf{h} + \mathbf{q} \tag{4.9}$$

Taking one of the terms, for example that in $\mathbf{a}_1$, the interference function then simplifies drastically:

$$\mathbf{a}_1 . \mathbf{Q} = \mathbf{a}_1 .\left(\mathbf{h} + \mathbf{q}\right) = \mathbf{a}_1 .\mathbf{h} + \mathbf{a}_1 .\mathbf{q} \tag{4.10}$$

The first term $\mathbf{a}_1 .\mathbf{h}$ is the Miller index component h, which is an integer. Thus in the interference function it becomes unity since $\exp(2\pi i) = 1$. This corresponds to strong Bragg diffraction when $\mathbf{q} = 0$. The second term when put in the interference function becomes

$$J_1 = \sum_{n_1=1}^{n_1} \exp\left(2\pi i\, n_1 q_1\right) = \frac{\sin\left(\pi\, n_1 q_1\right)}{\sin\left(\pi\, q_1\right)} \tag{4.11}$$

where q_1 is the component of $\mathbf{q}$ along the first axis in the diagram of Figure 4.1. This is in fact the first axis of the reciprocal lattice, which in cubic crystals is parallel to the real-space axes of the lattice. The final expression for intensity is

$$I = F^2 J^2 = F_{hkl}^2 \frac{\sin^2\left(\pi\, n_1 q_1\right)}{\sin^2\left(\pi\, q_1\right)} \frac{\sin^2\left(\pi\, n_2 q_2\right)}{\sin^2\left(\pi\, q_2\right)} \frac{\sin^2\left(\pi\, n_3 q_3\right)}{\sin^2\left(\pi\, q_3\right)} \tag{4.12}$$

At small values of the deviation vector $\mathbf{q}$ this reduces to the product of three functions of $\mathrm{sinc}^2(x)$ type. It is shown for one component in Figure 4.3.

Figure 4.3 brings out the following features of the scattering from a thin (or small) crystal:

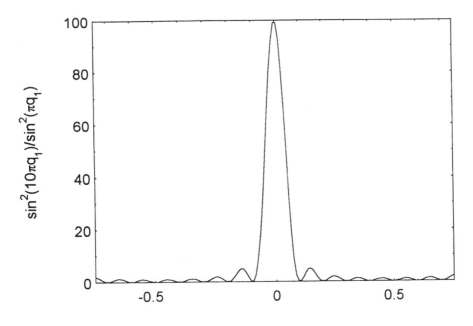

(a)

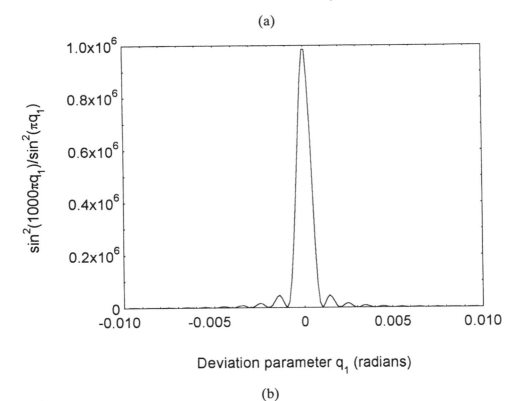

(b)

Figure 4.3 The scattered intensity as a function of one component of the deviation vector **q**. (a) $n_1 = 10$, (b) $n_1 = 1000$

1   The scattering is only intense near $\mathbf{q} = 0$, i.e. an exact Bragg reflection. The weak scattering from atoms is here reinforced by successive planes of atoms scattering in phase and strong intensity only occurs due to this reinforcement. Therefore, X-ray diffraction techniques integrate the scattering over many atomic layers.

2   The first zero occurs at $n_i q_i = 1$, i.e. the width of the diffraction peak varies inversely as the number of atoms. We expect appreciable broadening of diffraction spots for crystallites less than a few nanometres on edge and this is indeed observed.

3   The peak intensity, and also the integrated intensity, will be proportional to $|F_h|^2$.

4   The scattered intensity is proportional to the volume of the crystal. This implies that the scattering from a thin epitaxial layer, large in area compared with the beam diameter, will be proportional to the layer thickness.

5   For large single crystals extremely narrow rocking curves are predicted by the kinematical theory and these are not found. Dynamical theory is required for these cases.

4.2.4 *Anomalous dispersion*

When a phenomenon appears inconsistent with a simple theory, it often acquires the title 'anomalous', which persists long after the theory is properly understood. The behaviour of any mechanical or electrical oscillatory system has features near its resonance condition that are very different from those holding in the (much more usual) off-resonance conditions. For the scattering of X-rays, the resonant condition occurs near an X-ray absorption edge, when the energy of the incoming photon is just above that required to excite electrons between energy levels in the atom. We should expect to see more rapid variations than elsewhere in both the scattering and absorption behaviour, and these are termed *anomalous dispersion*. The treatment may be unified by notating the atomic scattering factors f_i (and structure factors that are derived from them) as complex numbers. The real part is the scattering that we have discussed so far, and the imaginary part is the absorption. Since we use exponential notation for the resulting wave, i.e.

$$A = \exp\left(2\pi i \phi\right) \tag{4.13}$$

where the complex function ϕ is linear with the atomic scattering factors, the imaginary part will drop out automatically into the usual absorption equation

$$I/I_0 = \exp\{-\mu t\} \tag{4.14}$$

as discussed in section 1.1. We take the real part to predict the scattering in, for example, a diffraction experiment. Near the absorption edge, and excluding all other effects such as Compton scattering, the real and imaginary parts of the scattering due to an oscillator must be related by a mathematical transformation known as the Kramers–Krönig transformation.

The effects of an absorption edge are shown in an important case for semiconductors in Figure 4.4. This shows the real and imaginary parts of the scattering factors for gallium and arsenic near the K absorption edges. The exchanges in values, and resultant intensities, that are shown can and have been used for composition-

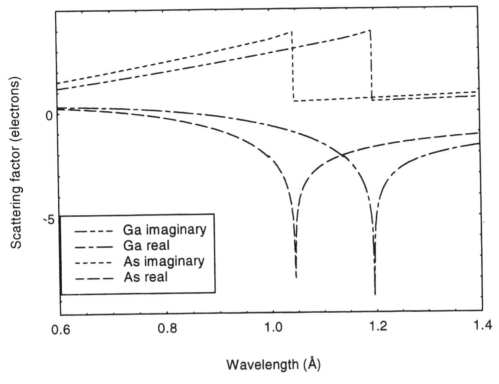

Figure 4.4 The real (scattering) and imaginary (absorption) parts of the atomic scattering factor for gallium and arsenic near the **K** absorption edges. After Cockerton *et al.*[1]

sensitive experiments such as the determination of stoichiometry in III–V semiconductors.[1] However, the tunability of a synchrotron radiation source is required for such experiments in most cases.

A further variant in the scattering near the absorption edge is the extended X-ray absorption fine structure (EXAFS). The energy absorbed by the material in exciting an electron from one energy level to another depends, in detail, on the atomic environment. This may be looked on alternatively as a back scattering and mutual interference of the emitted electrons, dependent on the environment, or as a modification of the state to which the ejected electron is excited, again dependent upon the environment. This is a powerful method of crystal structure analysis, especially for nanocrystalline or cluster regions in 'amorphous' materials.[2] It has only occasionally been combined with high resolution diffractometry and then only at synchrotron radiation sources.

4.2.5 *Reciprocal space geometry*

We surreptitiously introduced reciprocal space in Figure 4.1, in an attempt to show its usefulness before giving the formal definitions. It is so very helpful in the interpretation of many diffraction experiments that we need to understand it more fully.

We shall use it extensively in the discussion of triple-axis experiments, in which 'reciprocal space mapping' is an essential technique.

Figure 4.1 may be slightly extended to show a most useful construction, that of the Ewald sphere. Since we have elastic scattering, it is always true that the magnitudes of k_0 and k_h are both $1/\lambda$. A sphere of radius $1/\lambda$ can therefore define all possible incident and scattered beam vectors. The incident beam vector runs from the centre of this *Ewald sphere* to the origin, the scattered beam vector runs from the centre to any point on the surface of the sphere, and the scattering vector runs from the origin to the end of the scattered beam vector. Figure 4.5 shows a two-dimensional section of this three-dimensional construction. In this figure the scattering vector Q has been made to coincide with a vector h that satisfies the Bragg law (equation (4.9)) and we expect strong diffraction.

Of course, many reflections are possible from a regular lattice, and if the set of vectors such as h is represented then we can graphically visualise all the reflecting planes and consequent reflected directions in the crystal. We have deduced virtually all the rules already, but we formalise the description below.

1 The scale of reciprocal space is reciprocal length. (Note that in most physics texts this is scaled by a factor 2π for mathematical convenience in treatments of quantum theory. Here, and in general in crystallography texts, we use no such scaling.)

2 All directions in real space are preserved in reciprocal space.

3 A reciprocal lattice vector is constructed for each **plane** of the real-space lattice as follows:
   (a) the direction of the vector is perpendicular to the plane in real space,
   (b) the magnitude of the vector is the inverse of the interplanar spacing in real space.

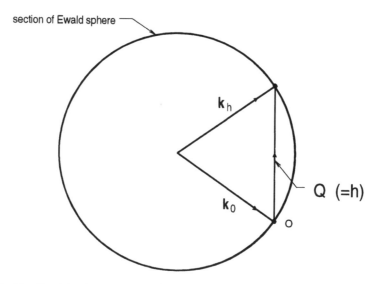

Figure 4.5 The Ewald sphere construction in reciprocal space

4   The end of each such vector, starting from the origin, is a reciprocal lattice point (sometimes abbreviated 'relp').

5   The reciprocal lattice is the set of relps.

Thus the reciprocal lattice axes are perpendicular to the (100), (010) and (001) planes in the real-space lattice. In cubic, tetragonal and orthorhombic crystals it is also true that they are parallel to the [100], [010] and [001] directions, but this is not true in other crystal classes. The general formulae for the reciprocal space axes, $\mathbf{a}_1^*$, $\mathbf{a}_2^*$ and $\mathbf{a}_3^*$ in terms of the real-space axes $\mathbf{a}_1^*$, $\mathbf{a}_2^*$ and $\mathbf{a}_3^*$ are

$$\mathbf{a}_1^* = \frac{\mathbf{a}_2 \times \mathbf{a}_3}{\mathbf{a}_1 \cdot [\mathbf{a}_2 \times \mathbf{a}_3]}, \quad \mathbf{a}_2^* = \frac{\mathbf{a}_3 \times \mathbf{a}_1}{\mathbf{a}_2 \cdot [\mathbf{a}_3 \times \mathbf{a}_1]} \quad \text{and} \quad \mathbf{a}_3^* = \frac{\mathbf{a}_1 \times \mathbf{a}_2}{\mathbf{a}_3 \cdot [\mathbf{a}_1 \times \mathbf{a}_2]} \tag{4.15}$$

Some examples will illustrate the concept. These are all drawn to scale for the silicon lattice and structure. In Figure 4.6 a high resolution experiment is shown. It is quite convenient to use reciprocal space to show the accessible reflections in a given experiment. These are shown in Figure 4.7 for a silicon specimen, with a (001) surface plane, and CuK$_\alpha$ radiation, for two of the orientations of the incident beam. The large semicircle contains all the points that are cut by the Ewald sphere as the incident beam is rotated 180° from just grazing the surface in one direction to just grazing the surface in the opposite direction. The small semicircle on the left contains points that cannot be accessed in reflection because the incident beam would enter from below the crystal surface. The small semicircle on the right likewise cannot be accessed in reflection as the diffracted beam exits through the crystal. These two regions are accessible in transmission. The reflection and transmission conditions are often called *Bragg case* and *Laue case* respectively.

In the powder diffraction experiment, a fixed wavelength and incident beam direction is used, and the random orientation of the powder grains means that the reciprocal lattice is rotated about all angles, centred on the origin. A diffracted beam ensues whenever a relp cuts the Ewald sphere. This is shown in two dimensions in Figure 4.8. This construction also makes it easy to see when reflections overlap, for example in Figure 4.9 we see the reciprocal space representation of the Laue back-reflection method. The Ewald sphere here sweeps through a range of wavelengths from the minimum given by the tube voltage (largest diameter Ewald sphere) to the maximum, which is controlled by absorption in the tube windows and in the air path. A crystal plane will diffract if its reciprocal lattice vector intersects the Ewald sphere at a wavelength within this range.

These examples show the great practical utility of the Ewald sphere construction. We did once hear Paul Ewald say, some sixty years after he laid the basis of X-ray scattering theory, that he wished people had named something else after him, as it was such a trivial idea!

4.3 Dynamical theory

As we have mentioned, the kinematical theory is unsatisfactory for predicting intensities of anything other than very thin or very small crystallites. The assumption that no energy is transferred into the diffracted beam might suffice for neutrino diffraction, but is very badly wrong for X-ray diffraction from most crystals of practical interest in the semiconductor industry. It is not merely the scaling factor

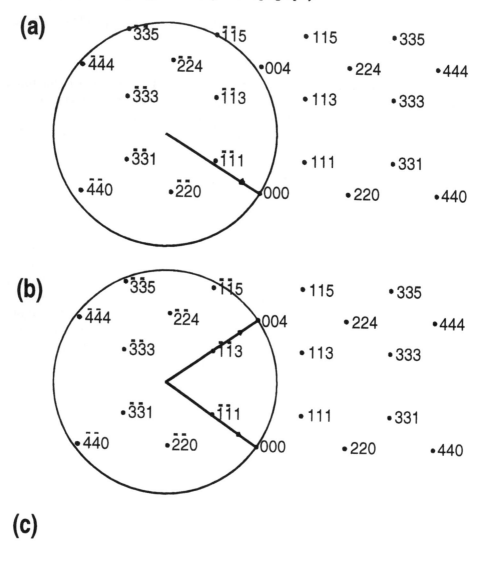

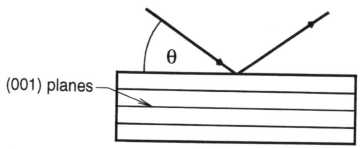

Figure 4.6 A high resolution experiment. In (a) the crystal is not yet aligned to the Bragg position and no diffracted beam occurs. In (b) the incident beam has been rotated so that the Ewald sphere falls on the 004 relp, and a diffracted beam ensues. The Ewald sphere is to scale for CuK$_\alpha$ and the reciprocal lattice is to scale for silicon

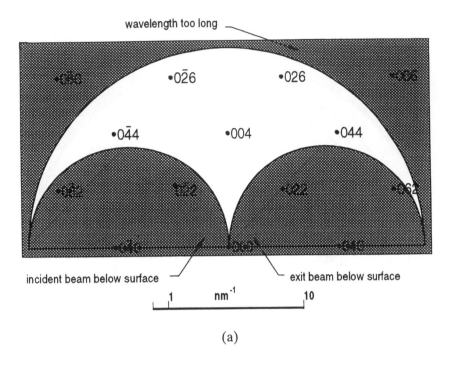

(a)

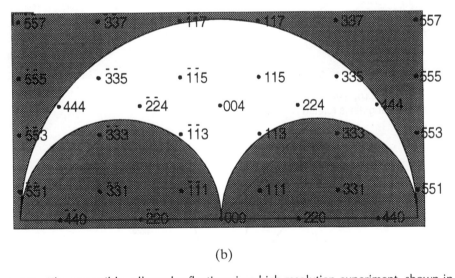

(b)

Figure 4.7 The accessible, allowed reflections in a high resolution experiment, shown in reciprocal space, for a silicon specimen with an (001) surface, using CuK$_\alpha$ radiation. (a) incident beam in (100) plane, reciprocal lattice section perpendicular to [100]. (b) Incident beam in (110) plane, reciprocal lattice section perpendicular to [1$\bar{1}$0]. Case (b) would occur if the incident beam were parallel or perpendicular to the plane of the (110) flat cut on many semiconductor wafers. Many more reflections are available in case (b). The semicircular segments showing accessible and inaccessible reflections are sections of hemispheres whose axes are vertical on this diagram

(a)

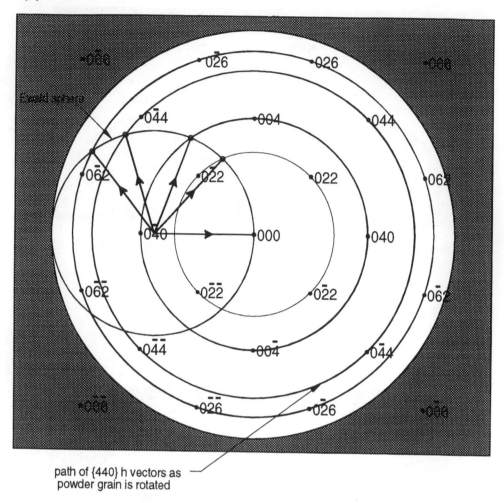

path of {440} h vectors as
powder grain is rotated

(b)

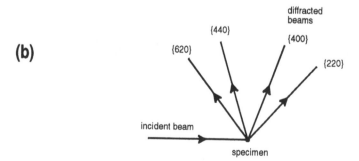

Figure 4.8 The powder diffraction experiment. (a) Reciprocal space notation. The Ewald sphere is fixed, and the lattice is rotated about all angles about the origin. Only the rotations about [100] are shown in this two-dimensional section. Intersections with the Ewald sphere define the diffracting conditions. (b) The corresponding diffracted beams in real space

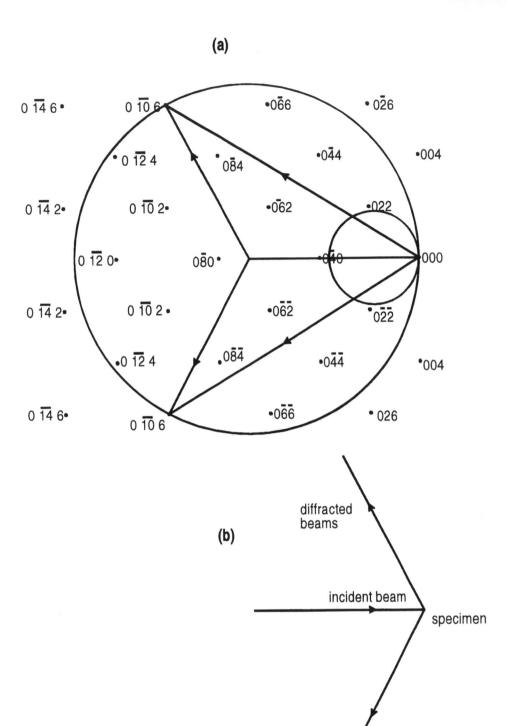

Figure 4.9 The Laue back-reflection method. The direction of the incident beam and the reciprocal lattice are fixed in space, and wavelengths are selected out of the beam. The Ewald sphere may be any diameter between the short and long wavelength cutoffs. The larger circle shows an intermediate wavelength and the diffracted beams that result at this wavelength

that is wrong. There is a saturation in the intensity as the crystal thickness increases and many additional interference effects caused by the presence of strong wave-fields that are not predicted by kinematical theory. The equations describing intensity as a function of thickness become seriously non-linear. However, thanks to theoretical studies over eighty years, the dynamical theory, which corrects these gross errors, is very well understood. Moreover, the recent arrival of packages providing numerical solutions on accessible personal computers means that industrial scientists can obtain precise, quantitative descriptions of the scattering that can be used in process development and quality control.

The simulation approach is based upon the idea that the diffracted beam from a set of reflecting planes is at the right angle to be rediffracted by the same planes. This is illustrated in Figure 4.10. It is seen that the energy is spread throughout a triangular region in this section known as the *Borrmann fan*. We expect this to result in a complicated expression for intensity, as indeed it does, though it can easily be implemented on a personal computer if the region is uniform. If it is non-uniform, as are all interesting industrial materials, this coupling of the diffracted and *forward-diffracted* beams (we can no longer think of the incident beam as transmitting unchanged through the crystal) must be treated locally around the inhomogeneities. The *Takagi–Taupin* theory is based upon the formulation and solution of a coupled pair of differential equations which represent the changes in amplitude in each of the forward-diffracted and diffracted directions.

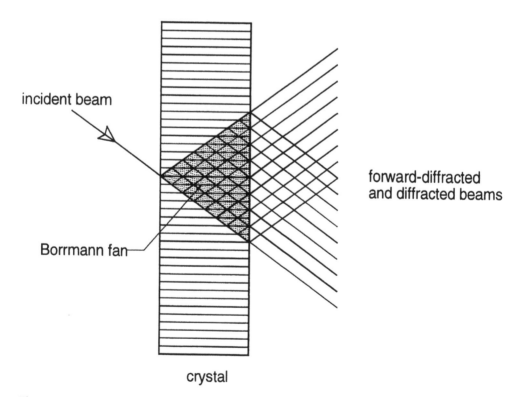

Figure 4.10 The diffraction and rediffraction of an X-ray beam from a set of reflecting planes. The triangle bounded by the incident beam and the diffracted beam from the entry surface is called the Borrmann fan

The Takagi–Taupin approach and the simulation methods that it allows are so powerful and useful that they are discussed in detail in the next chapter. However, in this approach we lose everything but the numerical amplitudes and intensities. We do not know how beams propagate through the crystal, how and when they are likely to interfere, what parameters control rocking curve width and shape; in other words the physics is buried. We therefore begin our discussion of dynamical theory by discussing the classical physical treatment and explaining the concepts that result. These will illuminate discussions of many aspects of experimental measurement and interpretation throughout the rest of the book. We shall not attempt to give the full derivations, since these are well treated in many excellent texts[3-11] but rather present the basis and assumptions of the theory and explain and interpret the conclusions and results. We largely follow the treatments of Batterman and Cole[6] and Hart.[7]

4.3.1 *Fundamental equations*

We are trying to discover what waves are generated inside and outside a crystal when it is illuminated by X-rays. The propagation of any electromagnetic waves in any medium is accurately described by Maxwell's equations. These are, in vector notation,

$$\nabla \times \mathbf{E} = -\frac{1}{c}\frac{\partial \mathbf{B}}{\partial t} \quad \text{and} \quad \nabla \times \mathbf{H} = \frac{1}{c}\frac{\partial \mathbf{D}}{\partial t} \tag{4.16}$$

where $\mathbf{E}$ is the electric field, $\mathbf{B}$ the magnetic induction, $\mathbf{D}$ the electric flux density, $\mathbf{H}$ the magnetic field and c the velocity of light *in vacuo*. This assumes that (at X-ray frequencies) the electric conductivity is zero and the magnetic permeability is unity, and is in Gaussian cgs units.

We want to introduce the properties of the crystal and of the X-rays and solve for the electric displacement or flux density, $\mathbf{D}$. Hart[7] gives a careful discussion of the polarisability of a crystal, showing that a sufficient model of the crystal for X-ray scattering is a Fourier sum of either the electron density or the electric susceptibility over all the reciprocal lattice vectors $\mathbf{h}$. Thus the crystal is represented as

$$\chi(\mathbf{r}) = \sum_{h} \chi_h \exp(-2\pi i \mathbf{h}.\mathbf{r}) \tag{4.17}$$

and χ_h can be put in terms of the now-familiar structure factor by

$$\chi_h = -\frac{r_e \lambda^2}{\pi V} F_h \tag{4.18}$$

where V is the volume of the unit cell over which the structure factor is calculated. The X-ray properties are introduced through the wavelength λ in this equation and we use the observed fact that the X-ray refractive index is very close to unity to make some simplifying assumptions, such as that χ is very small at these frequencies. Substituting in the Maxwell equations and eliminating the magnetic component, we find that

$$\nabla \times \nabla \times \mathbf{D} = -\frac{(1+\chi)}{c^2}\frac{\partial^2 \mathbf{D}}{\partial t^2} \tag{4.19}$$

What can we guess about the solution? It should clearly be a wave equation. We expect (from knowledge of Bragg diffraction) it to be a plane wave, or sums of plane waves. We use capital $\mathbf{K}_0$ and $\mathbf{K}_h$ for wavevectors inside the crystal to distinguish them from $\mathbf{k}_0$ and $\mathbf{k}_h$ outside the crystal. Inside the crystal the allowed wavevectors should satisfy conservation of momentum, that is

$$\mathbf{K}_0 + \mathbf{h} = \mathbf{K}_h \tag{4.20}$$

This *Laue condition* is a little less restrictive than the Bragg law, in that we no longer have the condition that $|\mathbf{K}_0| = |\mathbf{K}_h| = 1/\lambda$, but we still expect strong diffraction only when we are near the Bragg condition. Ewald proposed, and Bloch[12] showed that waves that exist in a crystal must have the periodicity of the lattice, that is, the solutions should look like

$$\mathbf{D} = \sum_h \mathbf{D}_h \exp\left(-2\pi i \, \mathbf{K}_h.\mathbf{r}\right) \tag{4.21}$$

with the $\mathbf{K}_h$ linked by equation (4.20). A qualitative argument in support of the Bloch theorem is that any waves that do not have this periodicity will be strongly damped by conflict with the 'electron density waves' that are the atoms themselves. We can further restrict ourselves to the case where two beams are very much stronger than any others. These are the diffracted and forward-diffracted beams shown in Figure 4.10. A glance at Figure 4.6 shows that it will require extremely careful adjustment to excite more than one diffracted wave with X-rays, so the two-beam assumption is very good. This is in contrast to the situation with electron diffraction in which the Ewald sphere is almost planar on the scale of this diagram and many beams are excited in almost all cases. The solution of the Maxwell equations inside the crystal (see for example Batterman and Cole)[6] is then expressed in terms of the amplitudes D_0 and D_h of these two beams, and is:

$$\left.\begin{array}{l} \left\{k^2\left(1+\chi_0\right) - \mathbf{K}_0.\mathbf{K}_0\right\}D_0 + k^2 C \chi_{\bar{h}} D_h = 0 \\ k^2 C \chi_h D_0 + \left\{k^2\left(1+\chi_0\right) - \mathbf{K}_h.\mathbf{K}_h\right\}D_h = 0 \end{array}\right\} \tag{4.22}$$

The polarisation factor, $C = \mathbf{D}_0.\mathbf{D}_h$. This is unity for σ polarisation, in which the electric field vector is perpendicular to the dispersion plane and $\cos(2\theta_B)$ for π polarisation in which the electric field vector lies in the dispersion plane. We may apply equation (4.20), which since $\mathbf{h}$ is real implies that $\mathrm{Im}(\mathbf{K}_0) = \mathrm{Im}(\mathbf{K}_h)$, and, in order to express the solution simply, introduce the very important idea of *deviation parameters*, α_0 and α_h. These express the deviation of the incident and diffracted wavevectors from the kinematic assumption where $|\mathbf{k}_0| = |\mathbf{k}_h| = 1/\lambda$, and are defined as

$$\left.\begin{array}{l} \alpha_0 = \dfrac{1}{2k}\left[\mathbf{K}_0.\mathbf{K}_0 - k^2\left(1+\chi_0\right)\right] \\[2mm] \alpha_h = \dfrac{1}{2k}\left[\mathbf{K}_h.\mathbf{K}_h - k^2\left(1+\chi_0\right)\right] \end{array}\right\} \tag{4.23}$$

In effect, the deviation parameters define the local refractive index of the crystal. The solution may now be expressed in terms of the relationship between the permissible deviation parameters

$$\alpha_0 \alpha_h = k^2 C^2 \chi_h \chi_{\bar{h}} \qquad (4.24)$$

This equation defines the geometry of diffraction in the two-beam dynamical theory. The amplitude ratio may also be expressed in terms of the deviation parameters as

$$\frac{D_h}{D_0} = \frac{2\alpha_0}{C\chi_{\bar{h}}k} = \frac{C\chi_h k}{2\alpha_h} \qquad (4.25)$$

and also that

$$\left(\frac{D_h}{D_0}\right)^2 = \frac{\alpha_0 \chi_h}{\alpha_h \chi_{\bar{h}}} \qquad (4.26)$$

These are the fundamental equations of two-beam dynamical theory, which allow us to predict the wavefields and their intensities inside (and outside) the crystal. Of course, there are many consequential derivations and solutions, especially when we apply the boundary conditions, which consist of the matching of waves so that they are continuous across the boundaries, and we shall quote such equations as the need arises. We shall first interpret these equations visually to demonstrate the principal qualitative conclusions.

4.3.2 *Dispersion space and geometry*

We have said that equation (4.24) gives us the allowable deviation parameters. Waves that satisfy this equation may exist inside the crystal for a given incident beam wavelength, others may not. We are going to draw a diagram that shows all the allowable wavevectors near the Bragg condition. Since equation (4.24) relates wavevectors to the local refractive index, it is a dispersion relationship. The diagram of all allowable wavevectors is therefore called a *dispersion surface*.

We should first correct the wavevector inside the crystal for the mean refractive index, by multiplying the wavevectors by the mean refractive index $(1 + \chi/2)$. This expression is derived from classical dispersion theory.[7] Equation (4.18) shows us that χ is negative, so the wavevector inside the crystal is shorter than that in vacuum (by a few parts in 10^6), in contrast to the behaviour of electrons or optical light. The locus of wavevectors that have this corrected value of k lie on spheres centred on the origin of the reciprocal lattice and at the end of the vector **h**, as shown in Figure 4.11 (only the circular sections of the spheres are seen in two dimensions). The spheres are in effect the kinematic dispersion surface, and indeed are perfectly correct when the wavevectors are far from the Bragg condition, since if $D_h \to 0$ then the deviation parameter $\alpha_h \to 0$ from equation (4.25). 'Far' means a few minutes of arc at most!

Indeed, Figure 4.11 itself is perfectly correct on the scale of the drawing. All the effects that the dynamical theory considers take place within the thickness of the lines at the intersection of the two circles at the point L_0. The shift in wavevector due to the mean refractive index is also undetectable on this drawing. Let us then magnify Figure 4.11 about a million times; the drawing is to scale and O and H are now about 50 km to the right. The circles centred on O and H are now indistinguishable from straight lines. The effect of the mean refractive index is shown as the shift

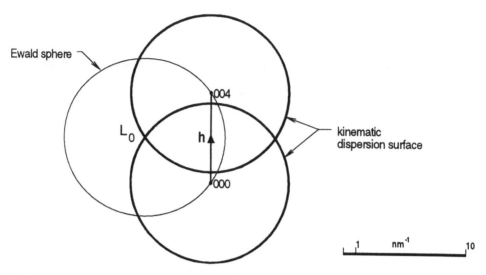

Figure 4.11 The kinematic dispersion surface. The circles centred on the origin O and the relp H, with radius $(1/\lambda_{vacuum})(1 + \chi/2)$ represent, in the plane shown, the allowable wavevectors in the crystal far from the diffracting condition. A section of the Ewald sphere is shown

from the intersection of these lines *in vacuo* at the *Laue point* L to the point L_0. However, equation (4.24) tells us that the deviation parameters, which are measured perpendicularly from these lines, can only be zero if the susceptibility is zero. In real cases, there must be a deviation, and its form, again from equation (4.24), is a hyperbola. The circles centred on O and H therefore do not represent solutions of Maxwell's equation when diffraction occurs. The hyperbola representing the locus of deviation parameters does. This is the *dispersion surface*.

For completeness we note that the dispersion surface is only strictly hyperbolic when the circles are accurately represented by straight lines. In three dimensions, of course, the lines become cones and the dispersion surface becomes a hyperboloid of revolution. We shall also have two independent dispersion surfaces for the two polarisations, σ and π. These must be taken into account in calculations with a randomly polarised source such as a laboratory generator, but we shall only draw one of the polarisations in the diagrams.

The dispersion surface has two branches, labelled 1 and 2. Waves from the two branches are in antiphase. This will be seen later to affect their absorption. Which of the allowable points (solutions of the Maxwell equation) that are selected depend on the incident angle and on the boundary conditions as discussed in the next section. One such point, labelled A, is shown in Figure 4.12. The values of the deviation parameters at A are shown. The vector from A to O is $\mathbf{K}_0$ and that from A to H is $\mathbf{K}_h$. Once these are fixed, we know the amplitude ratio from equation (4.25). This equation also tells us that the diffracted beam intensity will be a maximum when $\alpha_0 = \alpha_h$, at the centre of the dispersion surface. However, because of the displacement between L and L_0, the incident beam vector at this point is not quite parallel to the vector given by the conventional Bragg law calculation. There

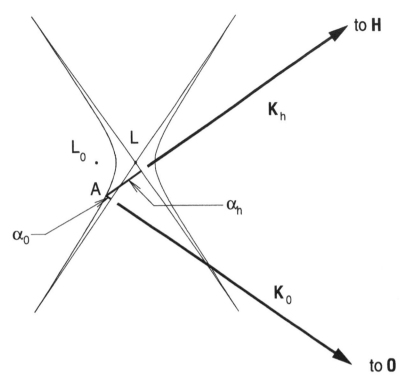

Figure 4.12 Magnification of the region near L_0 in Figure 4.11. L is the Lorentz point (the Laue point corrected for the mean refractive index). A is one of the tie-points selected on the dispersion surface, and α_0 and α_h are the deviation parameters at that point (shown on branch 2)

is a refractive index shift, always in the sense of increasing the apparent Bragg angle, usually around a few seconds of arc. The diameter of the dispersion surface is approximately $\chi_h k$, about $10^{-5}k$, and thus increases with the structure factor. We shall see later that this is directly proportional to the width of the rocking curve of the crystal.

It is qualitatively very helpful if instead of having to solve the equations or simulate the solutions every time, we can get some physical understanding which reduces the task. Ray optics is one such method; we look at the geometrical optics before the physical optics. Rays are reasonable concepts if diffraction broadening is not too great, that is, for a ray path $\approx \Lambda^2/\lambda$, where Λ is the width of the beam. For an X-ray beam 0.1 mm wide, this distance is about 1 mm, so the ray approximation does give a useful start. The direction of the ray is found by an interesting property of the dispersion surface. It is formally described by the Poynting vector, parallel to $\mathbf{E} \times \mathbf{H}$, and Kato[13,14] showed that this is perpendicular to the dispersion surface at the point where the amplitude ratio is defined. Thus, once the active points (solutions of Maxwell equations) are selected, the direction of energy flow or of rays is easily found. We shall find this very useful in X-ray topography in considering the contrast to be expected from defects at various depths in the crystal.

4.3.3 Boundary conditions

Any point on either branch of thé dispersion surface is an equally good solution of the Maxwell equations. However, the only points that will be selected are those that allow continuity of the wave – specifically of the electric displacement **D** – in all places, including the transition from outside to inside the crystal. The analogy with water waves may be useful. Here again there is an important dispersion relation as the velocity of waves in the deep ocean is higher than that in shallow coastal waters, and the resultant pile-ups are the surfer's delight. Nevertheless, the surface of the water is continuous through all these transitions.

Expressed formally, the wave must be matched in amplitude at the surface and in phase velocity parallel to the crystal surface. This implies that the tangential components of **D** and **H** must be continuous across the surface, and the components in the crystal surface of the wavevectors inside and outside the surface must be the same. If **n** is a unit vector normal to the crystal surface, whatever the values of $\mathbf{k}_0$ or the resulting Bloch wave inside the crystal, then

$$\mathbf{K}_0 - \mathbf{k}_0 = \delta \mathbf{n} \tag{4.27}$$

where δ is a scalar variable. This construction is shown diagrammatically in Figure 4.13. Just drop a line normal to the surface from the tail of the $\mathbf{k}_0$ wavevector, and the tails of the $\mathbf{K}_0$ wavevector(s) must lie somewhere on this line. But they must also lie on the dispersion surface. The intersection therefore gives the excited points, known as *tie-points*. Figure 4.14 shows the full construction for the Laue (transmission) case.

In the Laue case there are always two tie-points selected, one on each branch, labelled A and B. From each tie-point we shall generate wavevectors directed towards each of O and H. There are thus four wavevectors generated in the crystal for each polarisation, eight in all. The energy flow through the crystal is in the direction of the Poynting vector, that is, normal to the dispersion surface at the tie-points, and only at the exit surface do the waves split up into diffracted and forward-diffracted beams, as shown in Figure 4.15. We should repeat the boundary-condition construction for waves leaving the exit surface of the crystal, but it is identical to the entry-surface case if the surfaces are parallel. The displacement of the forward-diffracted beam from the ordinary attenuated direct beam is quite large enough to measure on a photographic plate, with suitable crystal thickness, and this formed one of the experimental tests of dynamical theory in highly perfect crystals.[15]

The Bragg (reflection) case is shown in Figure 4.16. Here the situation is a little less clear, because, according to the geometry, the normal from the surface intersects either two tie-points on the same branch of the dispersion surface or none at all. The Poynting vectors associated with the two tie-points are different; the energy flow from one point is directed into the crystal, but that from the other is directed outwards. The latter therefore does not generate any wavefields inside the crystal and can be ignored. Thus a single wavefield is generated for each polarisation.

In the cases where no tie-points are selected, no wavefields at all (other than a very rapidly decaying evanescent wave) are generated inside the crystal. The X-rays are effectively excluded from the crystal and the reflectivity, with a zero-absorption crystal, is 100%. This is the range of total reflection.

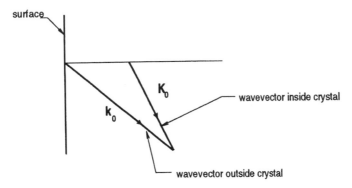

Figure 4.13 Matching of wavevectors across a boundary. The requirement that the component tangential to the surface is constant is satisfied by the above geometric construction

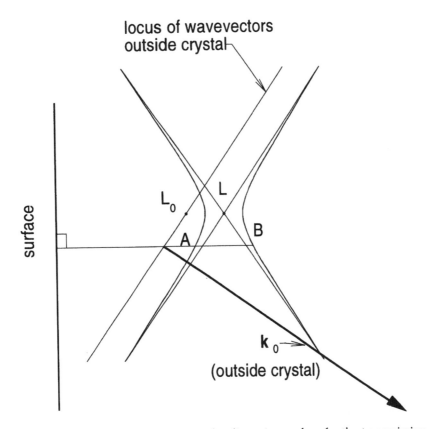

Figure 4.14 The selection of tie-points on the dispersion surface for the transmission (Laue) case, using the construction of Figure 4.13 (branch 2 is on the left, branch 1 is on the right)

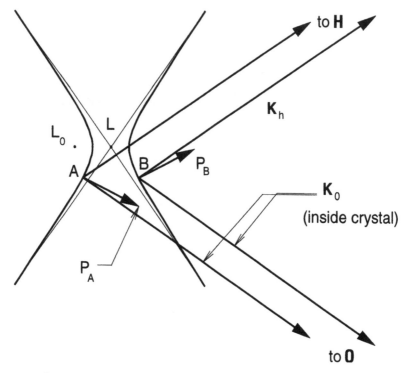

Figure 4.15 The energy flow through the crystal and the geometry of diffracted and forward-diffracted beams, for the case shown in Figure 4.14. P_A and P_B are the Poynting vectors associated with the tie-points A and B

4.3.4 Absorption

The above discussion has in effect been for materials with zero absorption, but this affects only the intensities. The construction of the dispersion surface and the wavevector matching are all performed on the real part of the wavevectors. When absorption is considered, the reflectivity in the Bragg case falls below 100% but it can still be over 99% for a low-absorption material such as silicon.

Far from the Laue condition the absorption shows the normal photoelectric absorption, as would be measured (with allowance for density) in a liquid or gas of the same atomic species. Close to the Laue condition, the absorption is quantified by the imaginary parts of the susceptibilities, leading to imaginary components of the wavevectors. These imaginary components are always normal to the crystal surface and hence the planes of constant attenuation are parallel to the surface. The attenuation coefficient $\mu(\mathbf{n})$ normal to the surface is given by

$$\mu(\mathbf{n}) = -4\pi \operatorname{Im}(\mathbf{K}_0)$$

(4.28)

It is more complex to express the attenuation along the direction of the Poynting vector,[16] so that we may see the absorption of 'rays' in the material, but it is interesting to quote one result, calculated by Batterman and Cole,[6] for 1 mm thick

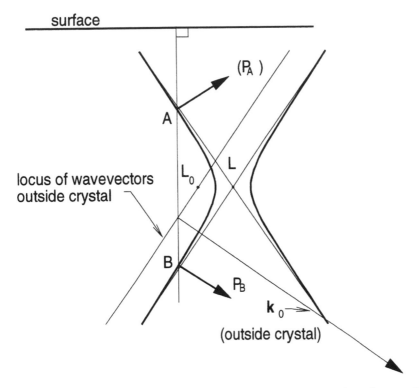

Figure 4.16 The selection of tie-points on the dispersion surface for the reflection (Bragg) case, using the construction of Figure 4.13

Table 4.2 Absorption (expressed as dimensionless μt values) for the various beams resulting from 220 diffraction in a 1 mm thick crystal of germanium

Branch	Polarisation	μt
2	σ	1.9
2	π	12.5
1	π	63.5
1	σ	74.0

germanium with CuK$_\alpha$ radiation. Far from the diffracting condition the normal photoelectric absorption gives $\mu t = 38$, thus the transmitted intensity would be exp{−38} or 3.14×10^{-14} of the incident intensity and quite undetectable. However, the two branches of the dispersion surface and the two polarisations show different absorptions when diffraction is important, as shown in Table 4.2 for 220 diffracting planes. The 2σ ray is 15% of the incident intensity and is easily measurable. This is known as the Borrmann effect or anomalous transmission. Again, it is not anomalous once the theory is understood, and it is very useful in the study of thick crystals by X-ray topography, since any defect in the Borrmann fan will disrupt the high transmission and give contrast in the image.

The values quoted in Table 4.2 are for symmetrical transmission; the 220 planes are normal to the surface, hence from the boundary-condition construction the tie-points excited are at the diameter of the dispersion surface. The Poynting vector is perpendicular to **h** and therefore the energy flow is along the Bragg planes. The differing absorption coefficients have a simple physical explanation. The $\mathbf{K}_0$ and $\mathbf{K}_h$ wavefields have components that travel in opposite directions, parallel and antiparallel to **h**, with wavevectors connected by the Laue equation. As with any other such pair of travelling waves they will set up standing waves in the crystal, normal to the Bragg planes, and it is straightforward to show that the periodicity in reciprocal space is just d, the interplanar spacing. It is also possible to determine the phase of the standing waves. For branch 2 waves, the nodes of the standing wavefield are on the atomic planes themselves whilst for branch 1 they are between the atomic planes, as illustrated in Figure 4.17. It is thus reasonable that photoelectric absorption, which is caused mainly by interaction with the inner K and L electron shells, should be a minimum when the antinodes of the standing wave (the regions of high electric field) fall between the atomic positions rather than on the atomic planes themselves. The interesting and useful point is the dramatic effect that this shows in the magnitude of the absorption.

4.3.5 *Penetration depth*

The penetration in the absence of absorption is governed by the extinction distance, ξ_h. This is the depth at which the intensity drops by a factor $1/e$ in a perfect crystal, and also the depth at which the Bloch wave in the crystal changes phase by a factor 2π.

$$\xi_h = \frac{\lambda \sqrt{|\gamma_h \gamma_o|}}{C\sqrt{\chi_h \chi_{\bar{h}}}} \tag{4.29}$$

Absorption effects combine with this to make the beam attenuate more rapidly. In the absence of extinction, the absorption depth t_a is given by

$$t_a = \frac{\sin \varphi}{\mu} \tag{4.30}$$

where μ is the linear absorption coefficient and φ is the angle between the incident beam and the crystal surface.

4.3.6 *Spherical waves*

As has already been emphasised, it is very difficult to approach a plane wave with the laboratory sources, which are properly described as spherical-wave sources. A single-crystal reflectivity profile from such a source will be dominated by the source profile rather than by the above formulae; the integrated intensity expressions will be multiplied by the source profile function.

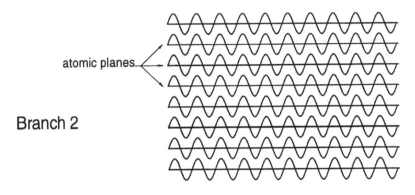

atomic planes

Branch 2

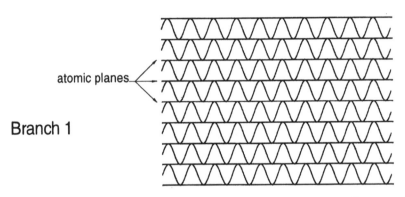

atomic planes

Branch 1

Figure 4.17 The standing wavefields set up in symmetrical Laue-case reflections. The nodes of wavefields from branch 2 of the dispersion surface lie on the atomic planes and the wavefields experience low absorption. The antinodes of wavefields from branch 1 of the dispersion surface lie on the atomic planes and the wavefields experience high absorption

The effect that a divergent source has inside the crystal is of great importance for X-ray topography, and it is useful to consider it qualitatively. Since the rocking curve profile corresponds to exciting tie-points on the dispersion surface sequentially over its hyperbolic range (Figure 4.18), it follows that a source with divergence greater than the rocking curve width will excite all the points on the dispersion surface within this range. A graphic way of putting it, due to Kato,[14] is that the whole dispersion surface is illuminated. As shown in Figure 4.18(b) this gives rise to energy flow within the Borrmann fan bounded by the incident and diffracted beam directions. This should be compared with Figure 4.10. This concept will be most useful in the interpretation of transmission X-ray topographs. Defects at any point within the Borrmann fan may contribute to the change of diffracted intensity at the exit surface of the crystal, and hence to image contrast.

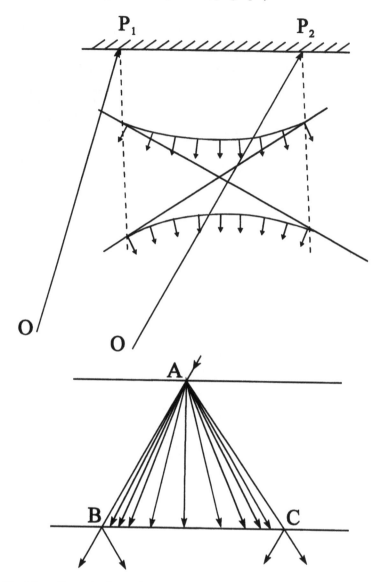

Figure 4.18 The effect of spherical incident waves on the excitation of Bloch waves. (a) Reciprocal space: the divergent incident beam has wavevectors ranging from P_1O to P_2O. (b) Real space: energy is distributed throughout the Borrmann fan ABC. The beams generated outside the crystal are indicated

4.4 Scattering in imperfect crystals

We will not treat in any detail in this chapter the more advanced topic of scattering of X-rays by distorted crystals. Appropriate theory will be introduced in later chapters as required. In this section we will merely give qualitative descriptions of the effects of various imperfections and quote some results.

4.4.1 *Thermal vibrations*

The atoms in a crystal are not in fixed positions but in continual motion due to thermal vibration. This has the effect of reducing the strength of the scattering. Zachariasen[9] shows that the effect on Laue–Bragg scattering is to replace the atomic scattering factors f_i by g_i, the latter being given by the expression

$$g_i = f_i e^{-M_k} \tag{4.31}$$

The exponent M_k depends on the mean square displacement of the atom from its equilibrium position and hence upon temperature. It is linear with $(kT/m)(\sin\theta/\lambda)^2$ where k is the Boltzmann constant, T the absolute temperature, θ the scattering angle, λ the wavelength and m the atomic mass (for a monatomic material). In addition there are complicated expressions dependent upon the crystal symmetry. As an example, for silicon at room temperature the f_i are reduced by approximately 6%. With this correction all the equations of dynamical theory still apply.

4.4.2 *Thermal diffuse scatter*

When the atoms are displaced from their lattice positions, they obviously spend part of the time in different positions. This means that they scatter X-rays in different directions from those predicted by the Bragg law. This is the thermal diffuse scatter. The maximum in the thermal diffuse scatter is displaced by typically a few minutes of arc from the Bragg positions, and is some six orders of magnitude weaker. In certain crystal structures, the time-averaged lattice vibrations may not have the same symmetry as the fundamental lattice and this will furthermore lead to diffuse Bragg reflections, which may appear between reciprocal lattice points of the fundamental structure.

4.4.3 *Mosaic structure*

Mosaic structure is the commonest defect in most real materials. A mosaic block is assumed to be a perfect crystal of a certain size, separated by an abrupt orientation change (larger than the angular width of the rocking curve) from adjacent blocks. These different blocks scatter independently. The effect on the scattering depends upon the scale of the mosaic blocks with respect to the extinction distance. If they are significantly larger than the extinction distance then dynamical scattering formulae apply to each block, and the incident beam is attenuated by this scattering before it reaches the next block. This is known as primary extinction. If they are significantly smaller than the extinction distance then kinematical scattering formulae apply and the beam is only attenuated by photoelectric absorption; this is secondary extinction. Extinction calculations for mosaic crystals are important in crystal structure analysis of imperfect crystals, but are less significant for high resolution diffractometry. We will have occasion to consider the scattering from mosaic regions, for example in reciprocal space mapping, but we rarely need to perform quantitative calculations on a three-dimensional assembly of mosaic crystals.

4.4.4 *Dislocation scatter*

One case that may be considered as mosaic-region scattering is the distorted region around an isolated dislocation. Two limiting cases are distinguishable. If the dislocations are well separated, that is, the kinematical scattering from a given dislocation is neglible when that from a neighbouring dislocation is significant, then the distorted volumes within the rocking curve width may be calculated and the contribution added.[17] If the dislocation strain fields overlap significantly, then a correlation function describing the averaged displacement fields must be constructed, from which the scattering may be calculated. With modern triple-axis diffraction equipment, the scattering from dislocation groups[18] and even from individual dislocations[19] is strong enough to measure, and quantitative information about dislocation density may be obtained.

4.5 Formulae for rocking curve widths, profiles and intensities

In this section we collect all the useful formulae for describing the shapes, widths and integrated areas or rocking curves.

There will only be a significant diffracted wave if both the deviation parameters are significantly greater than zero (equation (4.25)). A graphical way of looking at this is that this is the region in which the dispersion surface is hyperbolic, i.e. differing significantly from the circles centred on O and H. Figure 4.19 shows how the tie-points are selected in each case as the range of reflection is traversed. Note that tie-points on both branches are excited in the Laue case, but on one branch or the other in the Bragg case with, in the latter, a range of total reflection in which no tie-points are excited and the wave is excluded from the crystal.

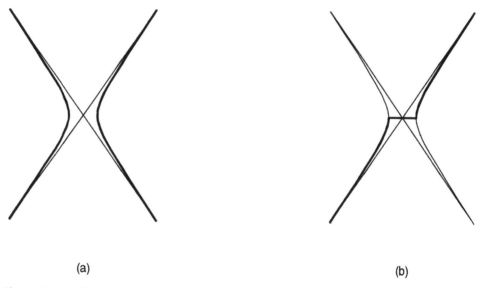

(a)   (b)

Figure 4.19 The range of strong diffraction in (a) the Laue case, (b) the Bragg case. The heavy lining shows the region of the dispersion surface excited as the rocking curve is traversed

In Figure 4.19 the hyperbolic region is shown separately to clarify its dependence on the diameter Λ of the dispersion surface. Since $\mathbf{K}_0$ is large, we may write

$$\Delta\theta_{1/2} = \frac{\Lambda}{\sin\theta_B K_0} \tag{4.32}$$

Since $h = 2K_0 \sin\theta_B$ (this is the Bragg law),

$$\Delta\theta_{1/2} = \frac{2\Lambda}{h} \tag{4.33}$$

for the symmetric case (reflecting planes parallel or perpendicular to the surface). In the asymmetric case the dispersion surface is 'viewed' obliquely, which will give a wider or narrower rocking curve for grazing incidence or grazing exit, respectively (Figure 4.20).

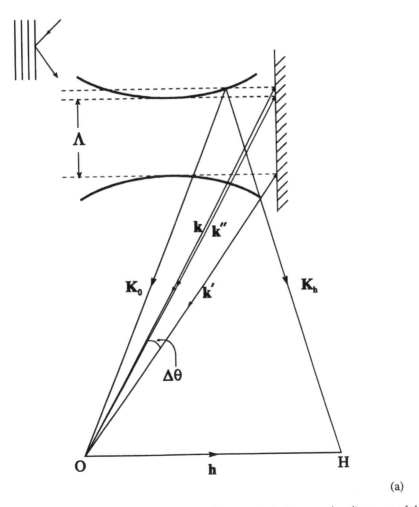

(a)

Figure 4.20 The dependence of the range of strong diffraction on the diameter of the dispersion surface. (a) Symmetric reflection. (b) Asymmetric reflection

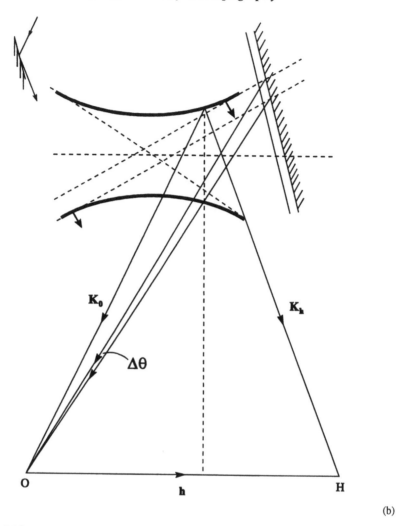

(b)

Figure 4.20 *(cont.)*

Equation (4.33) is at least approximately true for both transmission and reflection, but the details of the rocking curve are different in each case. The individual expressions for the profile are complicated to derive, and we shall simply quote the results. For generality the complete expressions for asymmetric diffraction are given, using the direction cosines γ_0 and γ_h. These are defined as the cosines of the angles made by the incident and diffracted beams, respectively, with the inward-going surface normals (Figure 4.21). These are general definitions, which may be simplified when the actual geometry is known. The asymmetry factor b is given by

$$b = \gamma_0/\gamma_h$$

where $b = 1$ for symmetric Laue case and $b = -1$ for symmetric Bragg case diffraction.

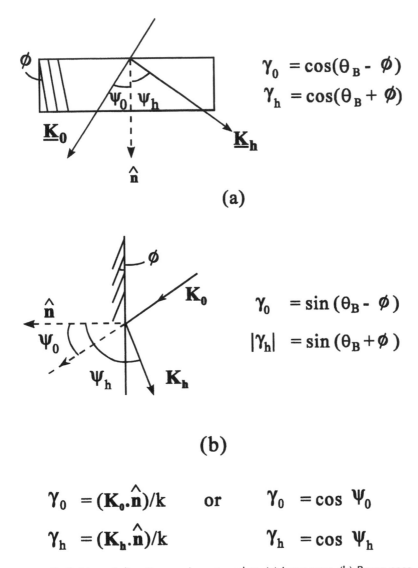

$$\gamma_0 = \cos(\theta_B - \phi)$$
$$\gamma_h = \cos(\theta_B + \phi)$$

(a)

$$\gamma_0 = \sin(\theta_B - \phi)$$
$$|\gamma_h| = \sin(\theta_B + \phi)$$

(b)

$$\gamma_0 = (\mathbf{K_0}.\hat{\mathbf{n}})/k \qquad \text{or} \qquad \gamma_0 = \cos\Psi_0$$
$$\gamma_h = (\mathbf{K_h}.\hat{\mathbf{n}})/k \qquad \qquad \gamma_h = \cos\Psi_h$$

Figure 4.21 Definition of direction cosines γ_0 and γ_h. (a) Laue case, (b) Bragg case

4.5.1 *Bragg case formulae*

The **rocking curve half-width** w (FWHM) is given by

$$w = \frac{2C\sqrt{\chi_h\chi_{\bar{h}}}}{\sin 2\theta_B}\sqrt{\frac{|\gamma_h|}{\gamma_0}} \tag{4.34}$$

Note that if w_s is the width of the rocking curve for the symmetric reflection,

$$w = w_s\sqrt{\frac{|\gamma_h|}{\gamma_0}} \tag{4.35}$$

The simplest expression for the **rocking curve profile** is in terms of a deviation parameter η, which varies from -1 to $+1$ in the range of total reflection. The reflectivity as a function of angle is then

$$R_B = |b| \left| \eta \pm \sqrt{\eta^2 - 1} \right|^2 \tag{4.36}$$

where η is defined as

$$\eta = \frac{-b\Delta\theta \sin 2\theta + \frac{1}{2} \chi_0 (1 - b)}{|C| |b|^{1/2} (\chi_h \chi_h)^{1/2}} \tag{4.37}$$

for the general asymmetric case in zero absorption (i.e. χ_0 and χ_h are real). The positive square root corresponds to tie points on branch 1 and the negative root to branch 2. Adding absorption we obtain real and imaginary parts of the deviation parameter:

$$\eta' = \frac{-\Delta\theta \sin 2\theta + \chi_0'}{|C| \chi_h'} \tag{4.38}$$

$$\eta'' = -\frac{\chi_h''}{\chi_h'} \left(\eta' - \frac{1}{|C|\varepsilon} \right) \tag{4.39}$$

where $\varepsilon = \chi_h'' / \chi_0''$.

Examples of rocking curves with low and high absorption will be shown in the next chapter.

The **integrated intensity** I_{hi} in the asymmetric Bragg case without absorption is

$$I_{hi} = \frac{4}{3} \frac{1 + |\cos 2\theta|}{\sin 2\theta} \sqrt{\left| \frac{\gamma_h}{\gamma_0} \right|} |\chi_h| \tag{4.40}$$

and with absorption is

$$I_{hi} = \frac{\lambda^2 (e^2/mc^2) N |F_H'| |P|}{\pi \sin 2\theta} \int \left(\frac{E_H^e}{E_0^e} \right)^2 d\eta \tag{4.41}$$

where

$$\left(\frac{E_H^e}{E_0^e} \right)^2 = G(\eta) - \left[G^2(\eta) - 1 \right]^{1/2} \tag{4.42}$$

$$G(\eta) = \frac{\eta^2 + X^2 + \left\{ \left[\eta^2 - (1 + Y^2 - Z^2) \right]^2 + \left[2X(\eta + Z) \right]^2 \right\}^{1/2}}{1 + B^2} \tag{4.43}$$

and

$$X = \frac{F_0''}{|P| F_H'}, \quad Y = \frac{F_H''}{F_H'}, \quad Z = \frac{|P| F_H''}{F_0''} \tag{4.44}$$

4.5.2 *Laue case formulae*

The Laue case **rocking curve width** in the limit of a thick crystal and zero absorption is

$$\Delta\theta = \frac{2|C|e^2\lambda^2|F_h'|}{\pi V mc^2 \sin 2\theta} = \frac{2|C||\chi_h|}{\sin 2\theta} \tag{4.45}$$

The **rocking curve profile** in the asymmetric Laue case without absorption is

$$R_0 = \frac{1+2\eta^2 + \cos 2\pi \dfrac{t}{\Lambda}\sqrt{1+\eta^2}}{2(1+\eta^2)}$$

$$R_h = \frac{1 - \cos 2\pi \dfrac{t}{\Lambda}\sqrt{1+\eta^2}}{2(1+\eta^2)} \tag{4.46}$$

where R_0 and R_h are the reflectivities of the forward and diffracted beams respectively.

Note that the intensity is now periodic with both the crystal thickness t and the reciprocal of the dispersion surface diameter, $1/\Lambda$. This is illustrated vividly in Figure 4.22, which shows the rocking curve as a function of crystal thickness.

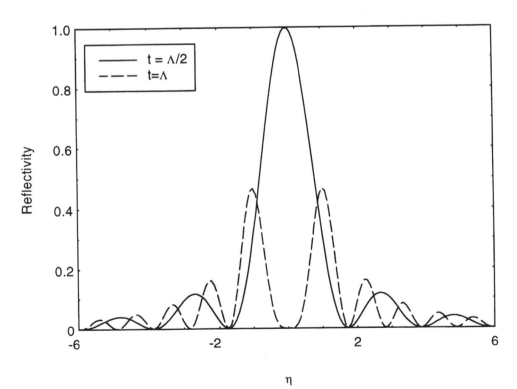

Figure 4.22 A plot of intensity as a function of deviation parameter and of crystal thickness

The **integrated intensity** in the Laue case without absorption is given by

$$I_{hi} = \frac{|C| |\chi_h|}{\sin 2\theta} \frac{\gamma_h}{\gamma_0} \int_0^{2\pi\Lambda_0^{-1}} J_0(z) dz \tag{4.47}$$

where J_0 is the zero-order Besel function.

Here is seen the most obvious contrast with the kinematical theory. Both curves are plotted as a function of thickness in Figure 4.23. It is seen that the formulae give similar results for small thicknesses, that the kinematic theory diverges drastically at larger thicknesses, and that the dynamic intensity shows oscillations about a saturated level after a thickness of about ξ_h.

With absorption the reflectivity is given by

$$R_H^{\eta'} = \exp(-\mu_0 t) \cos B \int_0^\infty \frac{\cos\left[\dfrac{B}{\left(1 + (\eta')^2\right)^{1/2}}\right]}{\left(1 + (\eta')^2\right) \cos B} \tag{4.48}$$

where

$$\eta' = \frac{\Delta\theta \sin 2\theta}{|C| |\chi_h'|} \tag{4.49}$$

and

$$B = \frac{\mu_0 t |C| F_h''}{F_0''} \tag{4.50}$$

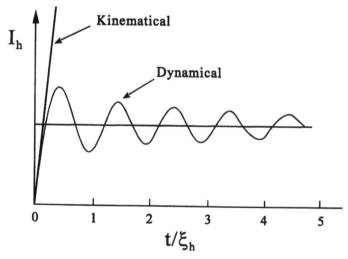

Figure 4.23 The transmitted integrated intensity as a function of thickness, as described by kinematical and dynamical theories

4.6 Summary

We have seen that we may obtain a quantitative description of X-ray propagation and scattering in perfect crystals. This theory is very well established and its accuracy has been tested very many times over more than 80 years. Crystals such as silicon and germanium are perfect enough to show quantitative agreement with the theory at least over seven or eight decades range of scattered intensity. The theory is routinely used for the design of X-ray optics and the modelling of scattering in epilayer crystals.

References

1. S. COCKERTON, G. S. GREEN & B. K. TANNER, Mater. Res. Soc. Symp. Proc., **139**, 65 (1989).
2. G. S. BROWN, in: Synchrotron radiation research, eds. H. WINICK & S. DONIACH (Plenum Press, New York 1980), p. 387; J. EVANS, in: Catalysis, Royal Society of Chemistry, Specialist Periodical Report 8, 1 (1989).
3. R. W. JAMES, The optical principles of the diffraction of X-rays (Ox Bow Press, Connecticut, 1982).
4. A. AUTHIER, in: Advances in structure research and diffraction methods, Vol. 3, eds. BRILL & MASON (1970) p. 1.
5. A. AUTHIER, in: X-ray and neutron dynamical diffraction: theory and applications, eds. A. AUTHIER, S. LAGOMARSINO & B. K. TANNER (Plenum Press, New York, 1997), p. 1.
6. B. W. BATTERMAN & H. COLE, Rev. Mod. Phys., **36**, 681 (1966).
7. M. HART, in: Characterization of crystal growth defects by X-ray methods, eds. B. K. TANNER & D. K. BOWEN (Plenum Press, New York, 1980) p. 216.
8. Z. G. PINSKER Dynamical scattering of X-rays in crystals (Springer, Berlin, 1977).
9. W. H. ZACHARIASEN, Theory of X-ray diffraction in crystals (Wiley–Dover reprint, New York, 1945).
10. M. HART, Rept. Prog. Phys., **34**, 435 (1971).
11. M. VON LAUE, Rontgenstrahlinterferenzen (Akad. Verlag, Frankfurt, 1960).
12. J. M. ZIMAN, Principles of the theory of solids (Cambridge University Press, London, 1964).
13. N. KATO, Acta Crystallogr., **11**, 885 (1958).
14. N. KATO, Acta Crystallogr., **13**, 349 (1960).
15. A. AUTHIER, Comptes Rendues Acad. Sci. Paris, **251**, 2003 (1960); A. AUTHIER, F. BALIBAR & Y. EPELBOIN, Phys. Stat. Sol., **41**, 225 (1970).
16. U. BONSE, Z. Phys., **177**, 385 (1964).
17. V. M. KAGANER, R. KOHLER, M. SCHMIDBAUER, R. OPITZ & B. JENICHEN, Phys. Rev. B, **55**, 1793 (1997).
18. E. KOPPENSTEINER, A. SCHUH, G. BAUER, V. HOLÝ, G. P. WATSON & E. A. FITZGERALD, J. Phys. D: Appl. Phys., **28**, A114 (1995).
19. M. S. GOORSKY, M. MEHKINPOUR, D. C. STREIT & T. R. BLOCK, J. Phys. D: Appl. Phys., **28**, A92 (1995).

5

Simulation of X-ray Diffraction Rocking Curves

In this chapter we discuss the computational implementation of the dynamical theory of diffraction: Takagi–Taupin theory, the solution for grids and thin layers, HRXRD simulation, deviation parameters, strategies, effect of strain, dislocation and defect arrays. We then give a number of examples of simulations.

5.1 Introduction

In Chapter 3 we went as far as we could in the interpretation of rocking curves of epitaxial layers directly from the features in the curves themselves. At the end of the chapter we noted the limitations of this straightforward, and largely geometrical, analysis. When interlayer interference effects dominate, as in very thin layers, closely matched layers or superlattices, the simple theory is quite inadequate. We must use a method theory based on the dynamical X-ray scattering theory, which was outlined in the previous chapter. In principle that formulation contains all that we need, since we now have the concepts and formulae for Bloch wave amplitude and propagation, the matching at interfaces and the interference effects.

Unfortunately, the classical dynamical theory, whilst greatly illuminating the physics of wave propagation, is not very convenient to use. There are many tedious calculations and, in particular, wave matching and beam multiplication problems to be handled at every interface. If the theory had remained in that state it is certain that high resolution X-ray diffractometry would not be the major tool that it is today. Ideally we want a theory that will take in our experimental rocking curves and give out the structure. That is not possible yet; it is one of the class of inverse problems with limited experimental information (in particular the phase of the X-rays and the limited sampling of reciprocal space which a rocking curve provides). What we do have is a practical theory that can be implemented on personal computers for simulating the rocking curve of a material whose structure is known. Comparison of the features and then the intensities of the simulated and experimental curves permits iterative refinement of the simulated structure. Simulation of rocking curves is an extremely powerful method of interpretation of complex structures. In addition, it is very valuable for the design of experiments, optimisation of data collection strategy and education of new researchers and operators in high resolution diffractometry.

5.1.1 The Takagi–Taupin generalised diffraction theory

This generalised diffraction theory, developed independently by Takagi[1,2] and Taupin[3] can be used to describe the passage of X-rays through a crystal with *any type* of lattice distortion. This theory is not formulated in terms of wavefields, but is a multiple scattering theory, familiar to electron microscopists as the Howie–Whelan equations. It assumes that X-rays are propagating as plane waves and that scattering is occurring both into and out of the diffracted beam. While this appears to be physically wrong, we must remember that Bloch waves are just a sum of plane waves in the forward and diffracted directions. Thus mathematically we can always keep the account straight by adding the correct phase factors into the scattering from forward to diffracted wave and vice versa.

The important thing about this theory for simulating the rocking curves of multiple and multilayers is that we assume a single wavevector and no matching at the boundaries needs to be done. So at the expense of understanding what physically goes on inside the crystal we have a mathematical calculating method which works splendidly. The theory can be applied equally well to deformed and distorted crystals as to a perfect crystal. It has therefore become the most powerful method both for interpreting rocking curves of complex epilayer systems and for simulating the contrast of defects such as dislocations in X-ray topographs. Using simulation techniques you can relate the scattered X-ray intensity to the microscopic lattice strains in the crystal.

The wavefield inside the crystal is described in a differential form; D_0 and D_h now are the total amplitude of the wave in the forward and diffracted beam directions, which may be slowly varying functions of position. How slowly? Surprisingly, the theory works very well for changes as abrupt as stacking faults and relaxed epilayer interfaces. K_0 and K_h are again the incident and scattered wavevectors. We also take $|K_0| = nk$, where n is the refractive index far from Bragg reflection. As is usual and accurate in most X-ray diffraction we use the two-beam approximation, that is, only the forward and diffracted beam wavefields have appreciable intensity. With these assumptions, Takagi and Taupin took a modified Bloch wave representation of the wavefield and obtained two coupled second-order partial differential equations expressed along the forward and diffracted beam directions s_0 and s_h (these are **unit** vectors in the directions of K_0 and K_h):

$$\frac{i\pi}{\lambda}\frac{\partial D_0}{\partial s_0} = \chi_0 D_0 + C\chi_{-h} D_h \tag{5.1}$$

$$\frac{i\pi}{\lambda}\frac{\partial D_h}{\partial s_h} = \left(\chi_0 - \alpha_h\right) D_h + C\chi_h D_0 \tag{5.2}$$

where C is the polarisation factor and α_h represents the deviation of the incident wave from the exact Bragg condition. This is a key parameter as it is this that we vary when we scan a specimen to collect the rocking curve. We treat mismatched layers, graded layers and all kinds of defects by their effect on a local distorted reciprocal lattice vector and hence on the deviation parameter.

For a perfect, uniform crystal, whether in bulk or as a thin layer, the Takagi–Taupin equations can be solved exactly as given in the next section. For the general case with multiple layers, however, it is necessary to integrate them numerically. The concepts of the dispersion surface are lost, and we cannot tell directly in which directions wavefields are propagating. They do give *directly* the intensities of the direct and diffracted beams emerging from the crystal, and all interference features are preserved.

5.2 Numerical solution procedures

All that we need to do now is to solve the Takagi–Taupin equations and plug in the deviation parameter to predict any intensity. For the general distorted crystal, numerical solution is used. The most common is that over a grid of points, which may be distorted in any way, for example to model the strains caused by a dislocation or precipitate at a given depth and hence to simulate the X-ray topographic image. This is discussed in Chapter 8. There have been many attempts to solve Takagi's equations analytically. The only solutions of real interest to us here are for the perfect crystal and the thin layer.

5.2.1 *Thin-layer and substrate solutions*

Halliwell, Juler and Norman obtained an important solution of the Takagi–Taupin equations for a uniform layer of known composition, structure and thickness.[4] This allows any one-dimensional strain distribution to be obtained by splitting up the crystal into lamellae of constant strain. The solution is expressed in terms of the variables

$$A = C\chi_{-h} \tag{5.3}$$

$$B = \frac{(1-b)\chi_0}{2} + \frac{\alpha_h \pi}{2} \tag{5.4}$$

$$D = \frac{\pi}{\lambda\gamma_0} \tag{5.5}$$

$$E = -Cb\gamma_h \tag{5.6}$$

where $b = \gamma_0/\gamma_h$. We also write

$$F = \sqrt{BB - EA} \tag{5.7}$$

These are all complex variables since the susceptibilities are complex. We obtain the amplitude ratio at the top (exit) of the layer, X, in terms of that at the bottom (entrance), X':

$$X = \frac{X'F + i(BX' + E)\tan(DF(z-w))}{F - i(AX' + B)\tan(DF(z-w))} \tag{5.8}$$

The variable z is the depth above the depth w at which the amplitude ratio is the known value X', in effect the thickness of the layer or lamella. We know the amplitude ratio deep inside the crystal; here, both the diffracted intensity and the incident amplitude are zero, but since their ratio must always be <1 then the amplitude ratio $X' = D_h/D_0(z)$ must also be zero. Using the above we may derive the reflectivity of a thick crystal,

$$X = -\left(\frac{B + F \operatorname{sign}(\operatorname{Im}(F))}{A} \right) \qquad (5.9)$$

The parameters A, B, D and E are of fundamental importance. They depend upon the crystal susceptibilities χ_0, χ_h and χ_{-h}, the cosines of the inclination angles of the incident and diffracted beams with the inward-going surface normals (γ_0 and γ_h, with the asymmetry factor $b = \gamma_0/\gamma_h$), the polarisation factor C, the wavelength λ and the deviation parameter α_h. Except for α_h, we may look up or calculate all of these from the composition and structure of the material and the geometry of the experiment. Thus the problem is reduced to the calculation of the deviation parameter.

5.2.2 Calculation of strains and mismatches

At reasonably large angles the deviation parameter is given by:[5]

$$\alpha_h = -2\Delta\theta_h \sin 2\theta_B \qquad (5.10)$$

where $\Delta\theta_h$ is the local deviation from the exact Bragg angle, taking account of lattice strains (with the sign convention that incident angles below the Bragg angle are negative) and θ_B is the local exact vacuum Bragg angle. At small angles, which are very important for analysing thin layers, a better expression is:[6]

$$\alpha_h = \left[\gamma_0 \left(\gamma_0 - \gamma_h - 2\sin\theta_B \cos\phi \right) + \frac{1}{2}\chi_0 (1-b) \right] \times \left[|b|^{1/2} C (\chi_h \chi_{\bar{h}})^{1/2} \right]^{-1} \qquad (5.11)$$

The calculation of the Bragg angle and the deviation parameter requires a rigorous procedure for dealing with strains and mismatches, whether caused by composition changes or (for example) ion implantation. A suitable method is as follows. All deviation angles are referred to the ideal Bragg angle of the substrate. The procedure for calculating the deviation parameter for an arbitrary layer is then:

1   Find the unit cell of layer material in the fully relaxed (i.e. bulk) condition from materials databases. For alloys of more than one material, Vegard's law is applied to the lattice parameters, Poisson ratios and structure factors. Find the susceptibility of the layers (the extremely small change in susceptibility when a layer is strained may be ignored).

2   Calculate the strains ε_{xx} and ε_{yy} that would be applied if the lattice parameters in the interface plane of the layer were forced to conform to the substrate (full coherent epitaxy). Multiply these by $(1 - R)$ where R is the (fractional) relaxation of the layer. (See the discussion of measurement of relaxation in Chapter 3.)

3   Calculate the layer strain normal to the interface, ε_{zz} from the relationship[7]

$$\varepsilon_{zz} = -\left(\varepsilon_{xx} + \varepsilon_{yy}\right)\left\{ \frac{v}{1-v} \right\} \qquad (5.12)$$

4   Apply the strains to the layer material, giving a new unit cell. Hence obtain the Bragg angle of the layer, the difference $\Delta\theta$ between the Bragg angle of the layer and that of the substrate, and the angle of tilt $\Delta\phi$ between the diffracting planes

in the layer relative to the substrate. The deviation parameter α_h is then calculated separately for the different diffraction geometries, as follows:

Symmetric case:

$$\alpha_h = \Delta\theta_s - \Delta\theta \qquad (5.13)$$

Asymmetric, glancing incidence:

$$\alpha_h = \Delta\theta_s - \left(\Delta\theta - \Delta\varphi\right) \qquad (5.14)$$

Asymmetric, glancing exit:

$$\alpha_h = \Delta\theta_s - \left(\Delta\theta + \Delta\varphi\right) \qquad (5.15)$$

where $\Delta\theta$, the deviation from Bragg angle in the substrate, is the controlled parameter, which forms the abscissa of the eventual rocking curve graph.

Note that the procedure in (3) is more general than the more usual approximate formula:

$$m = m^* \left\{ \frac{1-v}{1+v} \right\} \qquad (5.16)$$

m and m^* are the relaxed and X-ray mismatches. Note that the X-ray mismatch is not the strain in the layer but the mismatch between the substrate and the layer in the strained condition. This formula neglects second-order terms, which can become large for layers with large mismatch, and is less suitable for incorporating relaxation. The new procedure is much more general, allowing accurate calculation of asymmetric reflections and avoiding approximations in implementation at large values of the deviation parameter.

It is seen from the above equations that a tetragonal distortion is assumed in the layer. This is only strictly true if the substrate surface orientation is (001), though for symmetric reflections the treatment is valid for any orientation under the assumption of isotropic elasticity. However, the distortion would, for example, be trigonal on a (111), surface which would seriously affect the Bragg angle calculations.

5.2.3 *Calculation of reflectivity*

Thus we calculate the reflectivity of a whole layered material from the bottom up, using the amplitude ratio of the thick crystal as the input to the first lamella, the output of the first as the input to the second, and so on. At the top of the material the amplitude ratio is converted into intensity ratio. This calculation is repeated for each point on the rocking curve, corresponding to different deviations from the Bragg condition. This results in the plane wave reflectivity, appropriate for synchrotron radiation experiments and others with a highly collimated beam from the beam conditioner.

5.2.4 *Approximations and limitations*

The dynamical diffraction theory is very accurate and highly practical, but no theory provides a perfect description of the complexity of real specimens. What are the limitations and approximations, and how serious are they?

5.2.4.1 Two-beam approximation

Most practical simulation programs use only two beams, the forward and the forward-diffracted. That is, only one strong beam is present other than the incident beam. For most cases this is an excellent assumption, since it is quite difficult to set up a true multi-beam X-ray diffraction experiment. This is in contrast to the situation in transmission electron microscopy, where it is almost impossible to avoid multiple diffraction, and is caused by the longer wavelength (sharply curved Ewald sphere) used in most X-ray experiments. For very high energy diffraction, for example at advanced storage rings, multiple diffraction may become more significant but it may be ignored in ordinary characterisation experiments.

However, at grazing incidence angles near the critical angle the reflected beam becomes important, and eventually becomes dominant. This is an important effect and the two-beam case is seriously in error for angles of incidence close to the critical angle. Fortunately, the reflected beam decreases in intensity as the fourth power of the scattering angle, and is weak compared with the diffracted beam (close to a Bragg peak) above about two or three times the critical angle. The latter is 0.22° for silicon and 0.31° for GaAs, with CuK_α radiation and is linear with wavelength.

5.2.4.2 Vegard's law

It is common to approximate alloy behaviour in both lattice parameter and Poisson ratio by a linear interpolation between the end points of the alloy system, e.g. the values for $Ga_{0.5}Al_{0.5}As$ will be taken as half-way between those for GaAs and AlAs. In the absence of extensive experimental data, which are not available, this is the only practical general method. There is no general theory that describes departures from linear behaviour (known as Vegard's law, originally proposed[8] for the lattice parameter variations in well-behaved ionic crystals!). In systems that have been studied, the lattice parameter interpolation turns out to be reasonable. For example the silicon–germanium system departs only about 2% from Vegard's law for any mixture.[9] There is very little information available on the behaviour of the Poisson ratio with alloy content in semiconductors, but elastic constant behaviour is known to depend strongly on electronic interactions. It would not be surprising if there were large deviations from Vegard's law in certain pathological cases. It is of course impossible to estimate the magnitude of this error in ordinary cases without hard experimental evidence. Vegard's law errors are, of course, not in the X-ray scattering theory but in the data put in to model the structure. If good data are available this is not a source or error.

5.2.4.3 Dislocation broadening

At present this may only be modelled approximately, as discussed in Chapter 4, rather than fully incorporated into the dynamical theory. This is an important cause of peak broadening in 'difficult' materials such as strained layers.

5.2.4.4 Neglect of diffuse scatter

As also discussed in Chapter 4, dislocations also give rise to diffuse scatter from the near-core region. Point defects of all types also give diffuse scatter. Although we do not yet have complete theories to describe this at present, we can model the diffuse scatter quite well by a Lorentzian function of the form:

$$I = \frac{A}{k^2 + \Delta\theta^2}$$

(5.17)

In some cases, a higher power Lorentzian can be found to give a better fit.

5.2.4.5 Computational approximations

These are remarkably few in current commercial simulation packages (though they do differ in the extent to which they incorporate all the effects discussed in this chapter). The main sources of error are in the incorporation of the instrument function in which there is a significant trade-off between speed and accuracy for large data sets.

5.2.4.6 Overall accuracy and sensitivity

Although there is an uncertainty that we cannot quantify if Vegard's law is used, it is clear that other errors are below about 1% if the specimen is of good quality and is accurately described. The very large number of successful analyses by means of simulation, leading to successful operation of semiconductor devices that are critically dependent on alloy content, leads us to believe that Vegard's law errors are also at the per cent level. We therefore expect that for a reasonably good-quality specimen with dislocation density below about $10^3 \, cm^{-2}$ we should be able to obtain 2–5% agreement between simulation and experiment, for overall positions and intensities.

This is not to be confused with sensitivity. Certain features will show far higher sensitivity than this implies, as we shall see. The sensitivity is always enhanced for weak structural features if it can modulate the effects caused by strong features. Examples are, peaks that appear close to the substrate peak and modify its shape;[10] or very thin layers with large atomic number differences to adjacent layers.

5.3 Matching an experimental measurement

For laboratory-based systems the instrument function given by the effect of the beam conditioner must now be introduced. In Chapter 2 we discussed beam conditioners in detail and showed that they may be characterised in terms of an intensity which is a function of both divergence and wavelength.

5.3.1 Effect of beam conditioner

We first consider the beam divergence. As discussed in Chapter 2 we must perform a mathematical correlation (often miscalled convolution) between the plane wave rocking curves of the beam conditioner and of the sample crystals. If $R_1(\alpha,\lambda)$ and $R_2(\alpha,\lambda)$ are the reflectivities (in intensity) of the first and second crystals as functions of the angle of incidence α, at a given wavelength, λ. then the total double-crystal reflectivity $R(\beta,\lambda)$ at any angle β of the first crystal relative to the second crystal is given by[11]

$$R(\beta,\lambda) = \frac{\int\limits_{-\infty}^{\infty} R_1(\alpha,\lambda)R_2\{(\alpha-\beta),\lambda\}d\alpha}{\int\limits_{-\infty}^{\infty} R_1(\alpha,\lambda)d\alpha} \qquad (5.18)$$

The denominator (normalising constant) is the integrated reflectivity of the first crystal, and the λ dependence takes account of the spectral distribution of the source (see the next section). The integrals are strictly taken to infinity on either side, but have to be truncated to a reasonable value consistent with the accuracy required.

The second property of the beam conditioning system is the wavelength dispersion. When the beam conditioner is not the same crystal as the sample used in the non-dispersive geometry, we observe a broadening of the rocking curve as discussed in Chapter 2. Each wavelength will have its own independent rocking curve. In the non-dispersive $(+n,-n)$ geometry, or if the bandpass of the beam conditioner is very narrow, it is not necessary to include this effect. In the general case we must integrate the reflectivity at any point on the rocking curve over the wavelength:

$$R(\beta) = \int\limits_{\lambda_1}^{\lambda_2} R(\beta,\lambda)d\lambda \qquad (5.19)$$

Since the $R(\lambda)$ are all independent there is no normalising factor. The common case is the transmission of the the two K_α lines in the dispersive $(+n,-m)$ geometry. If the Bragg angles of specimen and reference are very different, then the $K_{\alpha2}$ and $K_{\alpha1}$ lines are separated and all peaks become doublets with height ratios $1:2$. To simulate this, it is adequate to simulate two rocking curves for the wavelengths corresponding to the $K_{\alpha1}$ and $K_{\alpha2}$ lines and add the results in the ratio of $1:2$.

However, if the dispersion is small, as in the case, for example, of GaAs and InP crystals mixed in the 004 reflections as reference and specimen, the rocking curve simply broadens. Then to a reasonable approximation, one can simulate this by adding a curvature equal to the angular separation of the $K_{\alpha1}$ and $K_{\alpha2}$ Bragg angles.

5.3.2 Effect of detector noise

A small addition to a logarithmic curve will be very evident at the lower end of the curve and virtually invisible at the top. For good matches with experimental curves plotted on a logarithmic scale (to see small peaks) it is necessary to add a background noise. With a well set up system, this is essentially the detector background, which may be measured at an angle remote from the substrate peak. An automatic normalisation of the simulated curve to the substrate peak value and the lower tail value of the experimental curve may also be useful. Figure 5.1 shows this effect.

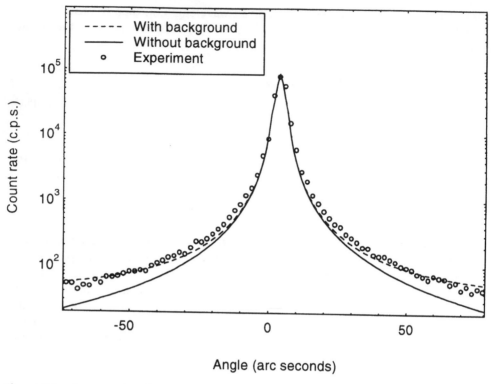

Figure 5.1 A comparison between experimental and simulated rocking curves with and without the inclusion of background. Si specimen 004 reflection CuK$_\alpha$ radiation

5.3.3 *Effect of sample curvature*

If the specimen or reference crystal is bent, and this can occur with wafers very easily if they are stuck onto the sample holder with too much adhesive, then the rocking curve is broadened. This occurs because different parts of the sample satisfy the diffraction condition at different angles. This may be simulated by correlating the rocking curve with a rectangle (in angle space) of width equal to the change in the incidence angle from one side of the beam to the other. The error introduced by assuming that the incident beam intensity is a rectangular function of angle is negligible. Figure 5.2 shows this effect.

5.4 **Limitations of the simulation approach**

The whole object of performing an X-ray scattering experiment is to be able to deduce the microscopic strains from the scattered intensity. However, except in a few special cases, the phase information is lost due to the fact that we can only measure intensity, not amplitude, of the scattered wave. Thus, direct determination of the structure by a Fourier-type transformation is not possible. Instead, we must adopt the method of simulating the X-ray scattering distribution from a model structure and comparing this with experiment. By sequential adjustment of the

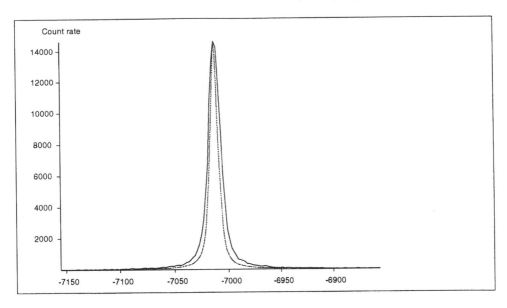

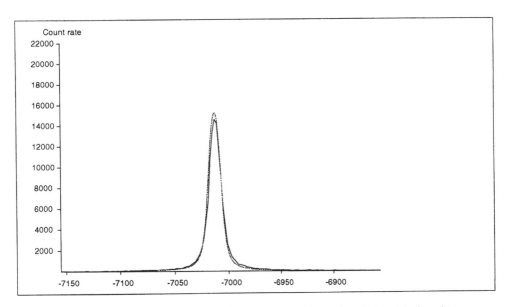

Figure 5.2 A (normalised) comparison between experimental and simulated rocking curves. InP specimen 004 reflection, CuK$_\alpha$ radiation. (a) Comparison without incorporation of curvature, (b) comparison with incorporation of curvature of 10 arc seconds. *x*-axis scale in arc seconds

structural parameters, we arrive at the best match to the experiment and in an act of faith, conclude that the structure of the material under investigation is that of the model. Notice that this method is adopted both in the interpretation of high resolution X-ray diffractometry data and X-ray diffraction topographs.

There are a number of unsatisfactory features about this procedure which it is important to examine. The first is the uniqueness of the solution. From a fundamental viewpoint, we may believe that the Uniqueness Theorem in electromagnetism suggests that there is indeed only one possible perfect match between experiment and simulation. However, even if this is the case, we can never have sufficiently perfect data for this stringent condition to be valid. All data are intrinsically statistically noisy, have a non-zero background and a finite range of wavevector covered. In practice, there can be no truly unique solution and this immediately leads to the second problem, that of local minima.

To obtain a match between experiment and simulation, we search for a global minimum in the difference between the two data sets. When many parameters are variable, such as in a superlattice structure, there appear a large number of local minima in this difference. Many of these can be very shallow and very difficult to leave; conversely the global minimum is often very deep but of very limited range in parameter space. In practice then, you need to be very close to the correct solution before the simulation will converge on experiment. At present, use of automatic fitting algorithms such as downhill Simplex is very hazardous, and while there are attempts to find algorithms which will identify local minima, a satisfactory solution to the problem appears to be some way off.

A third problem with simulation of high resolution diffraction data is that there is no unique instrument function. In the analysis of powder diffraction data, the instrument function can be defined, giving a characteristic shape to all diffraction peaks. Deconvolution of these peaks is therefore possible and fitting techniques such as that of Rietveld can be used to fit overlapping diffraction peaks. No such procedure is possible in high resolution diffraction as the shape of the rocking curve profile depends dramatically on specimen thickness and perfection. Unless you know the answer first, you cannot know the peak shape.

5.5 Examples of simulations

5.5.1 *Simulation strategies*

Extensive simulations take time; a typical, moderately complex material may require a few million floating-point complex multiplications to 20 decimal place accuracy! It is worth learning some effective time-saving strategies, which we suggest below.

- Decide what is the aim of the simulation before beginning. Is it to understand the specimen thoroughly, or to check that it is within process-control limits?
- Match the main features of the rocking curve first, using plane wave calculations and a single (sigma) polarisation. It should be possible to fit all the main peaks accurately in spacing and approximately in intensity. Then begin refinement to match the intensities and widths.
- Add the instrument function. If this is large it affects widths and relative peak intensities, but not integrated intensities.
- Add curvature (if physically reasonable). This has the same effect as the instrument function.
- Add the other polarisation state. This affects intensities by typically 20–30% but has little effect on widths. It doubles the time of computation.
- Note that instrument function and curvature corrections have a greater effect on narrow peaks (e.g. the substrate) than on broad peaks.

- Do not use too fine a step at the beginning. The computation of the plane wave rocking curve is linear with the number of steps, and with the number of layers. The correlation with the beam conditioner is independent of the number of layers but goes as the *square* of the number of steps or, if an integral transform method is used, with the logarithm of the number of steps.
- Note that a step that is comparable in size to the fringe spacing gives spurious periodicities – the aliasing effect, well known in control theory (Nyquist theorem) – and the final simulation must be performed with a fine step.

5.5.2 *An example simulation: deduction of material parameters*

For a general review of the use of simulation in high resolution diffraction see Tanner.[12] We now give an example simulation, showing the intermediate stages and our thought processes. We use a sequence of procedures which will improve the rate of convergence. The underlying philosophy is that, first, the **features** must be identified and matched and, only secondly, the **parameters** must be refined.

- **Use all the information at hand from the crystal grower**

This may seem elementary, but it is surprising how often one is asked to fit an X-ray rocking curve for a specimen of which almost nothing is known about the structure. Thus the first rule is to ask the crystal growers what structure they think they have grown. The second rule is not to believe them! Reliability in the answer can be assessed on several levels. It is unlikely that the grower has muddled InGaAs with AlGaAs and one can usually be confident in the sequence of compounds deposited. Exact compositions can be variable, but if other data such as photoluminescence or RHEED flux calibration are available, the reliability of your own crystal grower can be assessed over a period of time. Similarly, approximate thickness of various layers can be determined from other data, and always provides a starting point for trial simulations. Never, however, accept any assertion from a crystal grower as absolute truth. There is one splendid example of a commercial MBE kit on which the computer cannot count correctly the number of layers in a multilayer! It was some time before the computer screen was watched throughout the run and the evidence of the diffraction data accepted. In addition, interrogate your grower as to whether there are arbitrary capping or buffer layers present. These can lead to complex interference phenomena and the same degree of control over the deposition is desirable as over active layers. By taking care to establish a good working relationship with the crystal grower, a tight initial specification can be obtained. This will steer clear of many local minima and dramatically shorten time spent in simulation.

- **Initial measurements from the experimental data**

In Chapter 3 we discussed the extraction of data directly from the rocking curves. We will now apply these ideas and equations to help obtain our trial structure.

- **Identification and measurement of peaks**

Before attempting a first trial solution, it is essential to identify the Bragg peaks present in the rocking curve. A number of criteria may be applied.

117

(a) *The narrowest peak is almost always that associated with the substrate.* This rule works well for compound semiconductors such as AlGaAs on GaAs, CdHgTe on CdTe and InGaAsP on InP where the perfect crystal reflecting ranges are approximately the same for semi-infinite crystals of both substrate and layer materials. Where the rule will break down is for the situation where the substrate has a smaller structure factor than the layer material. For example, for a 1 μm thick layer of Si on GaAs, it is theoretically possible to have layer rocking curve halfwidth of 5 arcsec whereas the substrate halfwidth is 12 arcsec for plane wave CuK$_\alpha$ radiation in the 004 reflection. In practice, the Si will have relaxed by the creation of a large number of dislocations and the rocking curve width will have been broadened by the strains. (Further, it is unlikely that people will wish to grow Si on GaAs rather than vice versa, but the warning is there. Maybe diamond on silicon is a possibility which might arise in the future.)

(b) *Look at the lattice parameters to see whether you expect larger or smaller Bragg angles for layer or substrate.* Except when covering a heavily relaxed strained layer,[13] AlGaAs peaks are always at smaller Bragg angles than GaAs, as the AlAs lattice parameter is larger than that of GaAs. Where the lattice parameters of the binary components straddle that of the substrate, for example InAs and GaAs with respect to InP, there is no way of knowing *a priori* whether the substrate or layer peak will be at the higher angle. Measure the integrated intensity under the peaks. (Most plotting packages will do this if your data collection software does not.) Using Figure 3.9, make an estimate of the layer thickness from the ratio of the integrated intensities. If you regularly use another epitaxial system, it would be advisable to produce the equivalent figure by a series of simulations.

(c) *Measure the peak separation* of the various peaks and tabulate them. Use the equations of Chapter 3 to deduce obvious layer compositions, with the caveat that some peaks may be produced by strong interference effects rather than by actual layers.

(d) *If the structure is a superlattice, identify the equi-spaced superlattice satellite peaks and measure their separation.* If possible, determine the zero-order peak in the series of satellites. This correponds to the Bragg angle for the mean composition of the multilayer. If the zero-order peak can be identified, one further piece of data is obtained for the trial simulation. In the case of a binary superlattice, for example AlAs/GaAs, the position of the zero-order peak will give the well to barrier ratio. Together with the period, determined from the superlattice peak spacing, the thickness of both components is then known. For a ternary/binary system such as AlGaAs/GaAs, the thickness and composition cannot be determined uniquely.

Superlattice simulation is discussed in more detail in Chapter 6.

▪ Identification of fringe periods

A second feature which should be measured prior to a trial solution is the period of any fringe system seen in the rocking curve. This can be measured manually from the separation of adjacent peaks or by Fourier analysis of suitably treated data. If this latter course is taken, the Bragg peaks must be removed from the data by truncation of the angular range and the overall shape of the diffraction profile fitted and then subtracted to leave the periodic variation about a background of zero. Finally, autocorrelation of the data dramatically reduces the noise and sharpens up the Fourier spectrum. Procedures for undertaking such an analysis may be found in Hudson *et al.*[14] Except for the case where beating can be seen in the fringes, suggesting the presence of two closely spaced periods, this elaborate procedure is not normally necessary to a trial simulation. The equations for fringe spacing are given in section 3.4.8

5.5.3 *Example 1: single-layer structure*

We add the above deductions to the information from the crystal grower to obtain our starting trial structure. In this section, we use the above strategy to make a first attempt at simulation of the rocking curve shown in Figure 5.3, which we believe to be of a single layer of AlGaAs on GaAs.

We note:

(a) that the sharp peak on the right will be the GaAs peak,

(b) that the layer peak is separated by 120 seconds from the layer, thus suggesting a composition of $Al_{0.31}Ga_{0.69}As$ if the layer is totally strained,

(c) that the ratio of the integrated intensities is layer:substrate = 0.76:1, suggesting a layer thickness of 1.5 μm,

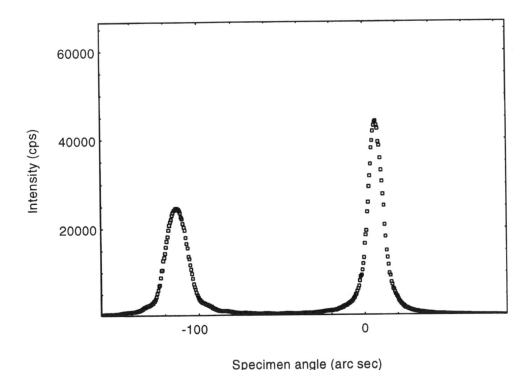

Figure 5.3 An experimental curve, consisting of a single layer of AlGaAs on GaAs

(d)   the interference period is 12.5 arc seconds also suggesting a layer thickness of 1.5 μm. (If these disagreed we would expect the fringe spacing to be a more reliable indicator of layer thickness.)

Figure 5.4 shows the fit between experiment and simulation on this first trial.

- **Parameter adjustment**

The fit in Figure 5.4 is not quite satisfactory, in that the position of the peak is displaced by 4 arc seconds in the simulation and the experiment. This is not surprising, as we assumed a Poisson ratio of 0.33 in estimating the composition. The fringe period is also too large, by about 10%. We should not be too worried by minor discrepancies such as these. Although the Poisson ratio of GaAs is well known at 0.281, that of AlAs is still in dispute and it is advisable to check the database of any commercial program to see what value is there (see Tanner *et al.*,[15] Goorsky *et al.*[16] and Bocchi *et al.*).[17]

From the peak splitting, we scale the composition in the ratio of 120/124 seconds and change the layer thickness to 1.6 μm. As seen in Figure 5.5, with this adjustment we get a good fit to both peak and fringe spacings.

The main difference between the experiment and simulation is now the broad background, peaking at the substrate Bragg angle. This is likely caused by diffuse scatter from dislocations and other defects in the substrate (and maybe in the layer too). We can model the diffuse scatter quite well by a Lorentzian function of the

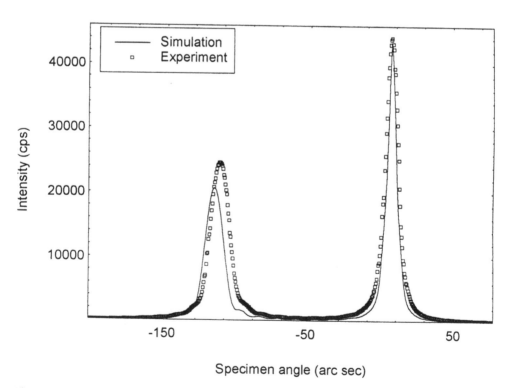

Figure 5.4   The results of the first simulation, with parameters extracted from the rocking curve

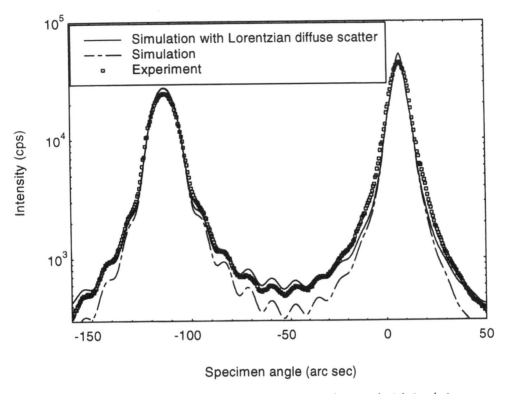

Figure 5.5 Comparison of rocking curve (small squares) with second trial simulation (dashed line). The effect of adding diffuse scatter is shown by the solid line

form given in equation (5.17). In this example, use of $k = 2.2 \times 10^3$ and $A = 1.2 \times 10^9$, with $\Delta\theta$ in arc seconds, gives a good match to the experimental data.

5.5.4 *A second example: a HEMT structure*

As a second example, we take the rocking curve from a high electron mobility transistor structure and show *exactly* all the stages in the simulation sequence. All rocking curves that were simulated when we first attempted the modelling are shown; this is a 'real-time' example!

The structure consists of a thin $In_xGa_{1-x}As$ layer grown epitaxially onto GaAs and covered with a GaAs cap layer. We know that the cap layer is about $1000\,\text{Å}$ thick and that the $In_xGa_{1-x}As$ layer is about $20\,\text{Å}$ thick. The experimental rocking curve was taken with $CuK_{\alpha 1}$ radiation in the symmetric 004 reflection, so we can immediately estimate the In composition assuming that the $In_xGa_{1-x}As$ is totally strained and unrelaxed. This is done simply from the relative position of the broad left-hand peak in Figure 5.6(a) with respect to the sharp substrate peak on the right. From

$$\Delta\theta = -\frac{\Delta d}{d}\tan\theta = -m * \tan\theta \qquad (5.20)$$

121

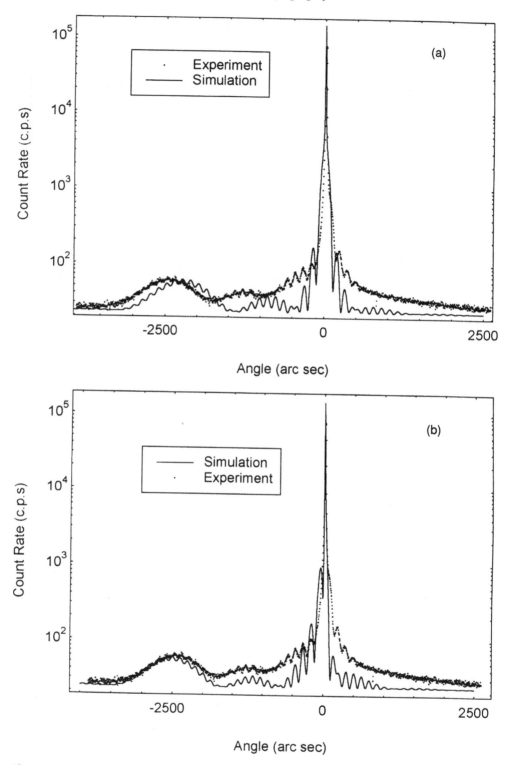

Figure 5.6 The HEMT structure and (a)–(f) the sequence of simulations as described in the text

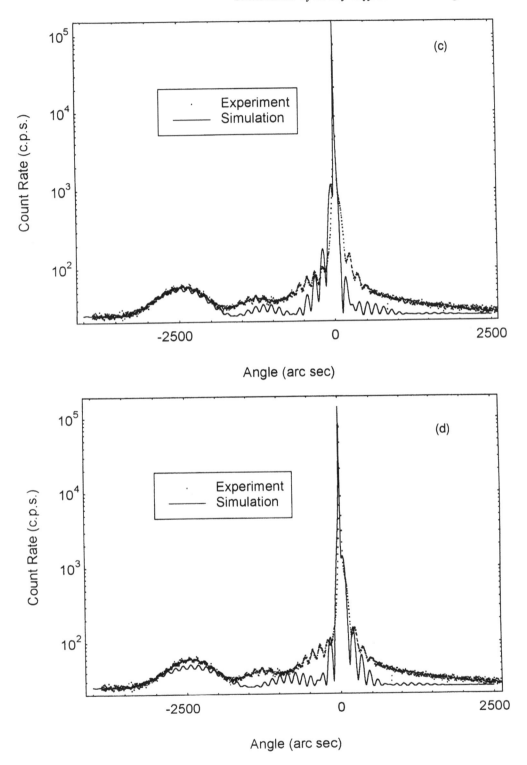

Figure 5.6 (*cont.*)

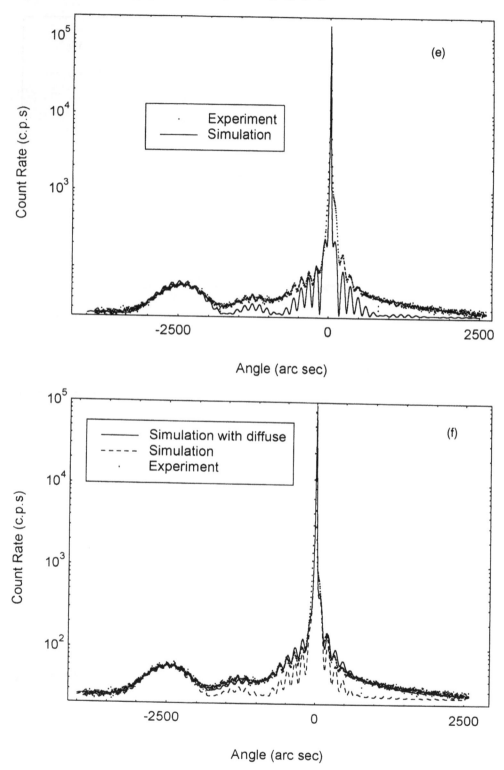

Figure 5.6 *(cont.)*

and as $\Delta\theta = 9600$ arcsec and $\theta = 33°$, we deduce an effective mismatch $m^* = 0.1432$. Assuming a Poisson ratio of 0.333, and thus $m^* = 2m$, the real mismatch m is 0.0716. With a(InAs) = 6.0584 Å and a(GaAs) = 5.65375 Å, we arrive immediately at a trial value of $x = 0.13$.

The high frequency fringe period corresponds to the combined thickness of both layers, or possibly the thickness of the capping layer. Assuming the latter, and measuring the fringe period $\Delta\theta_p$ to be 120 arcsec, gives from

$$\Delta\theta_p = \frac{\lambda}{2t\cos\theta} \qquad (5.21)$$

$t = 1580$ Å for the GaAs cap layer. The 'peak' at −946 arcsec from the substrate peak is in fact an interference fringe corresponding to the thickness of the InGaAs layer. Using the same formula, we deduce a starting thickness of 200 Å for the InGaAs layer.

The simulated curve in Figure 5.6(a) shows the result of using these parameters in the simulation. The simulated data have been normalised to 150 000 cps in the substrate peak, to allow for detector non-linearity. It is immediately clear that the layer peak is not in the correct position, and this arises from the fact that in deducing the starting parameters we assumed a Poisson ratio of 0.333. In the simulation, a correct value of the Poisson ratio has been used. The x value is then adjusted by the ratio of the simulated and experimental peak splittings, i.e. to (2526/2267) × 0.13 = 0.145. Figure 5.6(b) shows the result of using this value, and we see immediately that the peak positions now agree very well. They differ by 45 arcsec and Figure 5.6(c) shows a similar small adjustment in the same manner to $x = 0.1425$. Excellent agreement is now achieved.

However, the splitting of the interference peak at −946 arcsec is too small in the simulation by 140 arcsec, suggesting that the InGaAs layer is too thick in the simulation. Figure 5.6(d) shows the result of decreasing the InGaAs thickness to 170 Å and increasing the GaAs cap thickness to 1410 Å to compensate. This was clearly an error of logic as the effect was to move the interference peak closer to the substrate peak! Figure 5.6(e) shows the result of increasing the InGaAs thickness to 220 Å and leaving the total thickness at 1580 Å, as the short period interference fringe period is very well matched. This is *much* better!

The result is now sufficiently good to perform a convolution with the reference crystal. Note that the previous simulations were done under plane wave conditions, reducing the computing time significantly. The result shows no significant difference in the quality of the fit. What is now necessary is to adjust the InGaAs layer thickness to get the short period fringe position correct. Dropping the thickness to 215 Å shows that this is clearly the wrong way. A final alteration to 225 Å in Figure 5.6(f) gives almost perfect agreement.

Addition of a Lorentzian diffuse scatter term of the form $1.5 \times 10^7 / (\Delta\theta^2 + 3 \times 10^5)$ gives excellent agreement, as seen in Figure 5.6(f). Thus it has taken a total of only eight simulations to get a very good match to the data.

5.6 Summary

We have seen that the full dynamical theory may be solved conveniently for simple or layered materials that are uniform in the plane of the layer. The approximations are small, and accuracies of 2–5% are normal. Sub-nanometre sensitivity to layer thickness may be achieved in some cases, and simulation is the most powerful tool for detailed characterisation. An intelligent approach to simulation, using all available information, is shown to give rapid convergence to a satisfactory model.

References

1. S. TAKAGI, Acta Cryst., **15**, 1311 (1962).
2. S. TAKAGI, J. Phys. Soc. Japan, **26**, 1239 (1969).
3. D. TAUPIN, Bull. Soc. Fr. Mineral. Cristallogr., **87**, 469 (1964).
4. M. A. G. HALLIWELL, J. JULER & A. G. NORMAN, Microscopy of semiconducting materials, Inst. Phys. Conf. Ser., **67**, 365 (1983).
5. Z. G. PINSKER, Dynamical scattering of X-rays (Springer-Verlag, Berlin, 1978).
6. R. ZAUS, J. Appl. Cryst., **26**, 801 (1993).
7. L. D. LANDAU & E. M. LIFSHITZ, Elasticity (Pergamon, Oxford, 1972), pp. 13, 14, 55.
8. L. VEGARD, Z. Phys., **5**, 17 (1921).
9. A. Y. NIKULIN, A. W. STEVENSON & H. HASHIZUME, Phys. Rev., **B53**, 8277 (1996).
10. L. TAPFER & K. PLOOG, Phys. Rev. B, **40**, 9802 (1989).
11. R. W. JAMES, The optical principles of the diffraction of X-rays (Ox Bow Press, Connecticut, 1982) p. 308.
12. B. K. TANNER, J. Electrochemical Society, **136**, 3438 (1989).
13. B. K. TANNER, J. Phys. D.: Appl. Phys., **26**, A151 (1993).
14. J. M. HUDSON, B. K. TANNER & R. BLUNT, Adv. X-ray Analysis, **37**, 135 (1994).
15. B. K. TANNER, A. G. TURNBULL, C. R. STANLEY, A. H. KEAN & M. McELHINNEY, Appl. Phys. Lett., **59**, 2272 (1991).
16. M. GOORSKY, T. F. KEUCH, M. A. TISCHLER & R. M. POTEMSKI, Appl. Phys. Lett., **59**, 2269 (1991).
17. C. BOCCHI, C. FERRARI, P. FRANZOSI, A. BOSACCHI & S. FRANCHI, J. Crystal Growth, **132**, 427 (1993).

6

Analysis of Thin Films and Multiple Layers

In this chapter we discuss the double-axis rocking curves of thin single layers, the effects of thin interfacial layers and the Bragg case interferometer. We discuss characteristic rocking curves of superlattices and extraction of parameters, and show how the rocking curve builds up as more layers are added. We give strategies for analysis and recommend a starting point for simulations.

6.1 Review of rocking curves from thick layers

When the epitaxial layer thickness is quite high, typically of the order of one micrometre, we can apply the simple criteria discussed in Chapter 3 to determine the layer parameters from the rocking curve. The effective mismatch can be determined by direct measurement of the angular splitting of the substrate and layer peaks and the differential of the Bragg law. This simple analysis cannot be applied when the layer becomes thin, typically less than about $0.25\,\mu$m, where, even for a single layer, interference effects become extremely important. We consider these issues in section 6.2 below.

If a thick layer can be considered perfect, the integrated intensity can be used as a measure of the layer thickness, as described in Chapter 3. However, if the layer is thick and imperfect, the dynamical theory may not be appropriate. Then the integrated intensity is larger than that predicted and an overestimate will be made of the layer thickness. The criterion for a layer to behave as a perfect crystal is essentially that it must not contain inhomogeneous strains which give rise to effective misorientations greater than the perfect crystal reflecting range. Such tilts or dilations will lead to broadening of the rocking curve. An example of such a broadened rocking curve from a CdTe epitaxial layer grown on GaAs is shown in Figure 6.1. The simulated rocking curve from an ideally perfect layer has a rocking curve width of the order of 10 arc seconds, whereas the experimental curve is some hundreds of arc seconds in width. One approach is to assume that it is an ideally imperfect crystal consisting of many thin layers misoriented with respect to each other. The rocking curve can then be built up by adding the intensities of such layers convolved over the misorientation distribution function, in this case a Gaussian distribution. We may then expect that the integrated intensity will vary linearly with thickness and by use of the calibration table for the relative intensity of layer to substrate peak at very low layer thickness (Figure 6.2), we can deduce the thickness. It is difficult to specify the exact amount of broadening of the rocking curve necessary for this kinematical model to be appropriate as this is a function of the strain distribution through the layer.

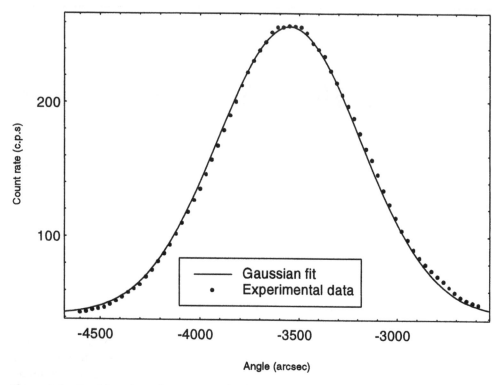

Figure 6.1 Double-axis rocking curve of CdTe on GaAs showing the broadening due to the very high dislocation density in the layer. (Courtesy R. I. Port, Durham University.) The rocking curve width in such thick, high mismatched layers falls with increasing layer thickness[1,2]

In this circumstance, it is more reliable to measure the absolute integrated intensity of the substrate peak and compare this with the integrated intensity from an equivalent crystal of the substrate material on which no layer has been grown. In the angular position for diffraction from the substrate, the layer will not diffract and the substrate peak intensity will be simply reduced by normal photoelectric absorption. For a symmetric reflection, it is easy to see that the integrated intensity I_s of the substrate peak with the layer of thickness t present is related to the integrated intensity of the bare substrate I_0 by

$$I_s = I_0 \exp\left[-\mu t / (2 \sin \theta_B)\right] \tag{6.1}$$

where μ is the linear absorption coefficient and θ_B is the Bragg angle. Provided that the layer composition is known, μ can be calculated. Note therefore that for ternary alloys, the Bragg peak splitting must be used to determine the composition and hence μ. When the reflection is not symmetric, and the Bragg planes make an angle φ with the surface, equation (6.1) is generalised to

$$I_s = I_0 \exp\left[-\mu t / \left\{\sin(\theta_B + \varphi) + \sin(\theta_B - \varphi)\right\}\right] \tag{6.2}$$

due to the different path lengths of the incident and diffracted beams. This approach is valid independently of the layer perfection.

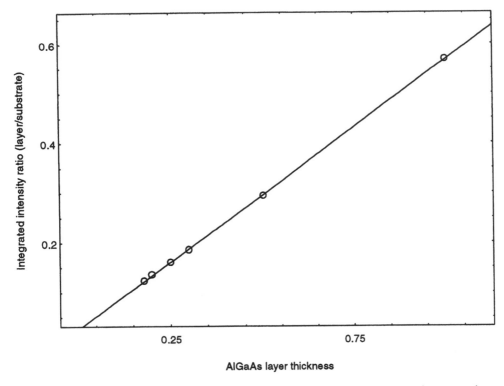

Figure 6.2 Ratio of integrated intensity of layer peak to substrate peak as a function of thickness for a thin AlGaAs layer on GaAs

6.2 Thin single layers

We have already seen something of the complexity and subtlety of the rocking curves which can arise when there are several thin layers present. This can be understood in terms of contributions from individual thin layers and thus it is important to have a clear picture of the effects of thin single layers.

The first point is one of simplicity, as the integrated intensity varies linearly with thickness when the layer is sufficiently thin to be in the kinematical diffraction limit. This is again valid irrespective of whether the layer is imperfect or perfect. It may be more convenient to measure the ratio of layer to substrate peak, which also varies linearly with thickness provided that the layer remains sufficiently thin for absorption to be neglected (Figure 6.2).

As discussed in Chapter 3, the interference (thickness) fringes can be used as a very direct method of layer thickness measurement. We note that the period is related to thickness (equation (3.32)) only by geometrical parameters and does not involve the structure factor. Thus, we might expect that for a layer of identical composition to the substrate, but with a different lattice parameter, the interference fringes will be present. This effect is seen very clearly in rocking curves of doped and ion implanted layers. What is not quite so obvious is that the interference fringes will be seen when the layers are of different composition but exactly lattice-matched.

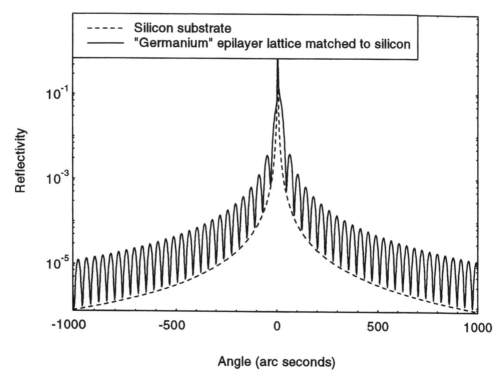

Figure 6.3 Rocking curve for a layer of 'germanium' with lattice parameter 5.43 Å, exactly lattice-matched to a silicon substrate. Interference fringes are seen

This effect is shown in the simulation of Figure 6.3, which is for the rocking curve from a layer of 'germanium' with a lattice parameter 5.43 Å on a silicon substrate. Although the example is somewhat artificial, the situation does occur in ternary alloys which are exactly lattice-matched to the substrate.

When the epitaxial layer is thin, typically below 0.25 μm, the differentiated Bragg law can no longer be used to determine the effective mismatch. As originally shown by Fewster and Curling[3] and subsequently studied in detail by Wie,[4] the peak splitting for a given composition varies as a function of layer thickness. As shown in Figure 6.4, the effect is negligible for layers of thickness greater than 0.25 μm, but varies very rapidly below 0.1 μm. The shift is dependent on the mismatch and is not related in any way to relaxation. It arises as a result of the interference terms in the expression for the intensity scattered from the thin layer, but a direct physical explanation has so far eluded the present authors. The effect is contained within the dynamical theory formalism of the rocking curve simulation programs based on the solution of the Takagi–Taupin equations, the data in Figure 6.4 being the peak splitting as calculated using the Bede Scientific RADS program.[5] Thus, in order to determine the mismatch in a thin layer, it is essential to use the simulation technique, or at the very least a pre-computed calibration curve. It is not a problem in cases where the mismatch is large, and is then often neglected.

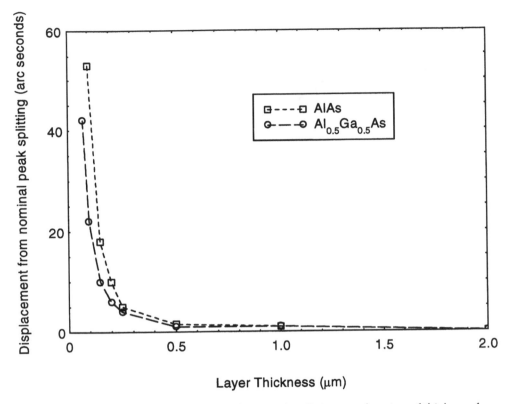

Figure 6.4 Variation in the substrate–epilayer peak splitting as a function of thickness for $Al_xGa_{1-x}As$ layers of constant composition ($x = 0.5$ and 1) on GaAs. Note that the absolute separation decreases with decreasing thickness. CuK_α 004 reflection

6.3 Graded thick layers

If the composition varies with depth through an epilayer, the rocking curve peak from the layer is characteristically broadened. The profile can be simulated by splitting the layer up into a sequence of thin lamellae of constant lattice parameter, which are assumed to scatter coherently (Figure 6.5). To a good first approximation, the scattering from the component lamellae is kinematical, and the rocking curve can be simulated by adding the amplitudes and phases from individual thin layers. The commercial simulation packages all permit thin layers of varying composition to be simulated in this way and, although it is within the dynamical theory formalism, the results are often identical to those of the kinematical theory.

When there is a linear grade in lattice mismatch between the top and bottom of the epilayer, the rocking curve has a characteristic wedge shape such as shown in Figure 6.6(a). For a quadratic dependence on depth, the curve becomes strongly asymmetric and falls more steadily to zero on one side of the peak (Figure 6.6(b)). The interference fringes can be quite complex and are not easy to analyse qualitatively. The period is neither associated with the thickness chosen for the component lamellae nor characteristic of the total thickness. As seen in Figure 6.6(c), the interference fringe spacing changes on going from a linearly graded

131

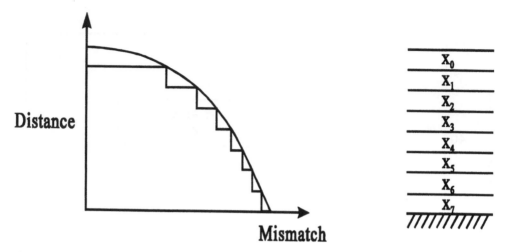

Figure 6.5 The approximation of a continuously varying lattice parameter into a series of lamellae of constant lattice parameter

structure to a quadratically graded structure. Although the end points are the same, here 0% aluminium to 100% aluminium in $Al_xGa_{1-x}As$, not only does the shape of the curve change, but so too does the interference fringe character. These fringes are genuine and not simulation artefacts. Experimental examples can be found in the literature.

Such rocking curves were found very commonly in the early days of development of quaternary systems for long-haul telecommunications links, and epilayers grown by chemical vapour epitaxy showed the effects particularly well.[6] Current III–V layers do not often show the effects, but this is not true for II–VI compounds containing mercury, such as the important infrared detector material $Hg_xCd_{1-x}Te$. Figure 6.7 shows an example of a rocking curve from a thick layer of another mercury compound, $Mn_xHg_{1-x}Te$, grown by metalorganic vapour phase epitaxy on a CdTe buffer layer on GaAs. The simulated rocking curve assumes that the layer is divided into 20 lamellae, each differing in the Hg concentration by $x = 0.05$. Such Hg segregation is also found in $Hg_xCd_{1-x}Te$ devices used for infrared sensors.

There are examples in the literature where there have been random fluctuations in the composition through the depth. Such curves are almost impossible to simulate in detail, but typical characteristics have been simulated by Lyons.[8]

6.4 Bragg case interferometers

In Chapter 5, we have already seen an example of the rocking curve from a high electron mobility transistor structure, where a thin, highly mismatched layer is situated between two identical layers. Figure 6.8 shows another such rocking curve, this time for an $Al_{0.6}Ga_{0.4}As$ layer between two $Al_{0.3}Ga_{0.7}As$ layers on a GaAs substrate. Although the peak from the thin $Al_{0.6}Ga_{0.4}As$ layer is itself of low intensity, the presence of this layer has a dramatic impact on the two $Al_{0.3}Ga_{0.7}As$ layers. The latter peak is modulated very strongly and almost appears to be split. As the

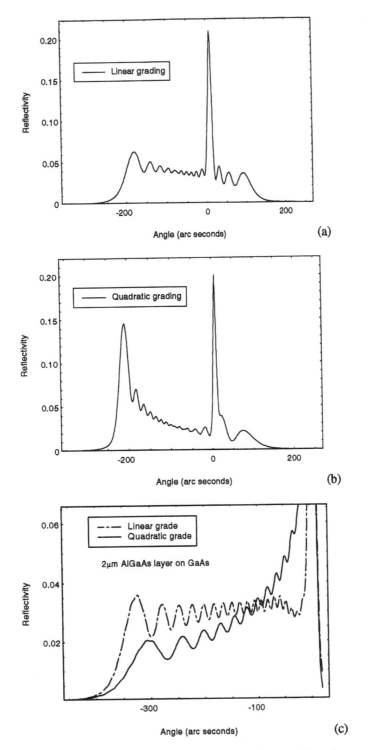

Figure 6.6 Simulated rocking curves of In$_x$Ga$_{1-x}$As on InP for: (a) a linearly varying composition with distance, $x = 0.0067t + 0.525$ where t is the distance in micrometres from the substrate interface; (b) a quadratic dependence on distance from the substrate, $x = -0.022t^2 + 0.0133t + 0.525$, total layer thickness 3 μm, 004 reflection, CuK$_\alpha$ radiation, InP reference crystal; (c) 2 μm thick Al$_x$Ga$_{1-x}$As layer on GaAs varying linearly and quadratically from $x = 0$ to $x = 1$

133

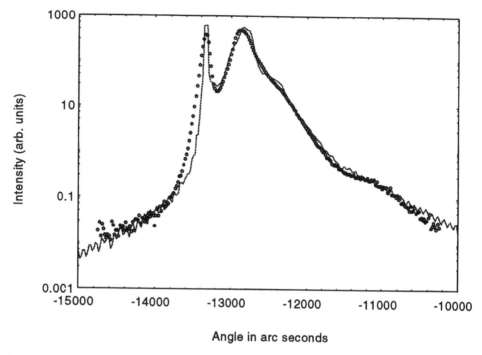

Figure 6.7 Experimental and simulated rocking curves of an $Mn_xHg_{1-x}Te$ grown on a CdTe buffer layer on GaAs. Slit collimated θ–2θ scan. (Courtesy C. D. Moore)[7]

sandwiched layer thickness is changed, the position of the peaks changes, altering the relative heights of the main and subsidiary peaks from the $Al_{0.3}Ga_{0.7}As$ layers. This example is a particular case of a general class of epilayer systems which behave as Bragg case interferometers.

The Laue (transmission) case X-ray interferometer was invented by and characterised in detail by Bonse and Hart in the mid-1960s.[9] Recent applications have been reviewed in detail by one of the present authors in the 1996 Erice school proceedings.[10] The realisation that the 'ABA' epilayer structure behaved like a Bragg case X-ray interferometer led to a very direct physical picture of the origin of the interference fringes in the rocking curves.[11] In Holloway's analysis, he pointed out that if the B layer had the same lattice parameter as the A layers, then provided the B layer was thin, the rocking curve profile would be independent of the B layer thickness as it simply inserted an integer number of Bragg planes into the structure. The phase difference between waves scattered from the top and bottom A type layers was then only changed by a multiple of 2π, leaving the rocking curve unaltered. If there was a mismatch m^* between A and B type layers, then the phase was altered by an additional amount given by $2\pi m^* t/d$ where t is the B layer thickness and d is the Bragg plane spacing. For a layer of thickness d/m^*, the interference fringes in the rocking curve fringes move along the angular axis by one period. The rocking curves for layers whose thickness increases by integer multiple of d/m^* are, to first order, identical.

Thus the sensitivity of the rocking curve to the presence of thin buried layers can be very high if the mismatch is high. Tapfer and Ploog,[12] Green *et al.*[13] and more

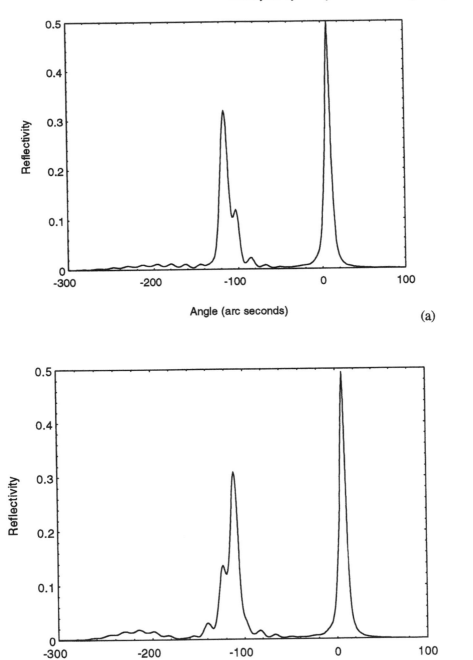

(a)

(b)

Figure 6.8 Rocking curve from a layer of $Al_{0.6}Ga_{0.4}As$ sandwiched between two 1 μm $Al_{0.3}Ga_{0.7}As$ layers on a GaAs substrate. Note the interference fringes and the change in phase of the fringes with respect to the layer peak as the AlGaAs layer thickness is changed. $Al_{0.6}Ga_{0.4}As$ layer thickness (a) 0.2 μm, (b) 0.3 μm

recently Li *et al.*[14] have used this phenomenon to deduce the presence of a thin interfacial layer with a high mismatch. For high mismatch (e.g. $In_{0.2}Ga_{0.8}As$ and GaAs) the sensitivity is at a single atomic layer. Note, however, that if the mismatch is low as in Figure 6.8, the thickness sensitivity is low and changes in thickness of several hundred angstroms may be necessary to observe effects.

6.5 Superlattices

Modern high-speed and optoelectronic devices regularly contain periodic sequences of layers of alternating composition. By use of such a period superlattice of small repeat layer thickness, the band structure of semiconductors can be dramatically altered and such multiquantum wells can exhibit a wide range of novel phenomena. As the effective bandgaps can be tailored by varying the thickness of the quantum well, this provides a very flexible design tool for the electronic engineer. Additionally, it has been found that superlattice structures can be used to prevent mismatch dislocations propagating, thereby providing a means of eliminating relaxation, which would lead to the presence of defects which reduce unacceptably the lifetime of light-emitting devices.

The rocking curves from such one-dimensional 'artificial crystals' are quite striking and a typical example is shown in Figure 6.9, where we see a sequence of

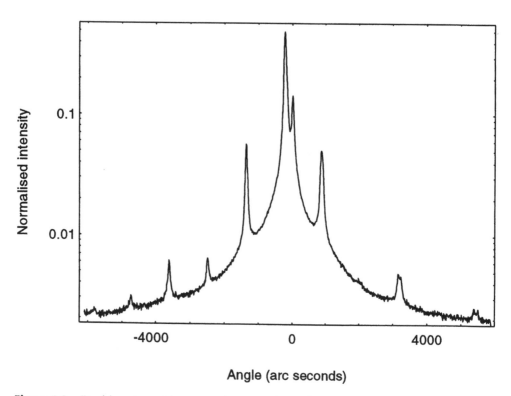

Figure 6.9 Double-axis rocking curve from an AlGaAs/GaAs superlattice CuK$_\alpha$ radiation, 004 reflection

equispaced satellite peaks associated with the superlattice of AlGaAs and GaAs. We can understand the origin of the superlattice satellites by noting that the artificial periodic structure will have a corresponding set of points in reciprocal space. The period in real space is large and therefore the separation of reciprocal lattice points in reciprocal space is very small. Additionally, it is in one dimension only and thus we have a line of reciprocal lattice points such as shown in Figure 6.10. The important point to note is that all reciprocal lattice points in the basic crystal, including the 000 point (which corresponds to grazing incidence reflectivity measurements) are modulated by the superlattice. The origin of the satellites then becomes clear and in triple-axis reciprocal space maps (Chapter 7) the reciprocal space structure is clearly seen.

Superlattice structures containing only a few repeats can be difficult to analyse and result in some quite unexpected effects. Consider for example an epilayer of total thickness $1\,\mu m$ which is composed of two layers L and M, each $0.5\,\mu m$ thickness and mismatched from the substrate by -600 and -1200 ppm respectively. The rocking curve (Figure 6.11) is very clearly composed of two layer peaks, one corresponding to each layer. However, if the total thickness is held the same, but the epilayer divided up into four layers each of $0.25\,\mu m$ thickness, then we have six peaks. Although the two strongest peaks are close, there is no peak which corresponds to the composition! Differentiation of the Bragg equation would suggest that there are six layers present, each of different composition. Further subdivision results in five peaks for six layers and, on division into ten layers of thickness $0.1\,\mu m$, we find a main peak appearing at the average composition of the structure. This is the so-called zero order peak.

The positions of the subsidiary peaks can be recognised as being determined by the thickness of the layers, or in another form, the superlattice period. These are the satellite peaks which we described earlier. It is best seen in a similar example to Figure 6.12, but with a larger mismatch between layers and plotted on a logarithmic scale. Here we show the simulated curves for an $In_xGa_{1-x}As$ structure on InP, of total epilayer thickness of $1\,\mu m$. The $In_xGa_{1-x}As$ is composed firstly of two layers of composition $x = 0.5$ and $x = 0.43$, each of thickness $0.5\,\mu m$. On subdivision into four

(a)

Substrate

(b)

Figure 6.10 (a) Real-space superlattice structure and (b) the corresponding structure in reciprocal space

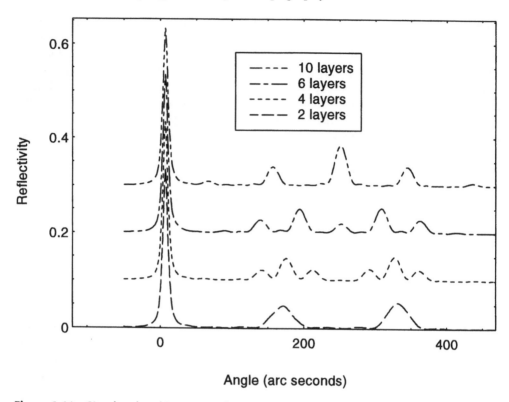

Figure 6.11 Simulated rocking curves from an epilayer of total layer thickness 1 μm, subdivided into 2, 4, 6 and 10 layers of alternate composition. The curves are displaced for clarity

alternating layers of composition $x = 0.5, 0.43, 0.5, 0.43$, each of thickness 0.25 μm, we see that the layer peaks are strongly modulated by the thickness fringes. Note that this modulation splits the peaks, leaving a minimum at the position corresponding to the differentiation of Bragg's law. Also note from the middle curve in Figure 6.12 that alternate fringes have been enhanced in intensity. Subdivision into eight layers (four repeats of $x = 0.5$ and $x = 0.43$ layers) results in the appearance of clear satellite peaks, the separation of which is determined by the sum of the pair of $x = 0.5$ and $x = 0.43$ layers. While this is confusing on a linear scale, examination of the data on a logarithmic scale makes the origin of the individual features clear.

It is worth noting that for superlattices where the two components differ only by the substitution on one of the zincblende structure sites (e.g. $Al_xGa_{1-x}As$/GaAs) it may be advantageous to use quasi-forbidden reflections such as 002. In such reflections, the intensity is determined by the difference in the atomic scattering factor of the atoms on the two zincblende sites, for example $|f_{Ga} - f_{As}|$ for the case of GaAs. For $Al_xGa_{1-x}As$, the intensity becomes $|(1-x)f_{Ga} + xf_{Al} - f_{As}|$. There is thus a large change in scattering factor between the two components and excellent contrast of the superlattice satellites is obtained. However, the penalty is that the overall intensity is down and, depending on the signal-to-noise ratio inherent in the experimental arrangement, it may be better to use the lower contrast but higher signal of

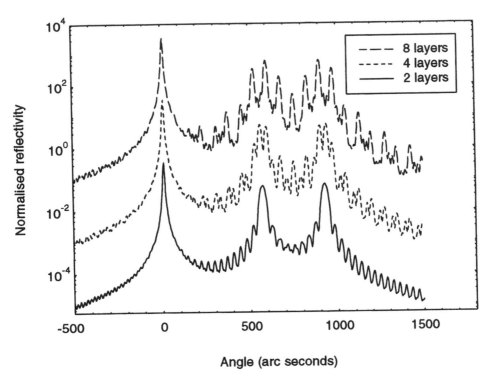

Figure 6.12 Simulated rocking curve from sequence of layers of total thickness 1 μm, divided into 2, 4 and 8 repeats of In$_x$Ga$_{1-x}$As layers of composition $x = 0.5$ and $x = 0.43$. The curves are displaced for clarity

the allowed 004 reflection rather than the higher contrast but lower signal of the quasi-forbidden 002 reflection.

6.6 General characteristics of large repeat superlattices

Device engineers and crystal growers are concerned with determining the following parameters of a superlattice structure:

- the spatial period of the structure
- the thickness of the repeating unit
- the compositions of the layers
- the dispersion in the repeating period
- the interface roughness
- the interface grading

These parameters can be found from the rocking curve, with the exception that roughness cannot at present be distinguished from grading if only double-axis diffraction is used.

Although the rocking curve from a superlattice or multiquantum well structure may be quite complex, there are a number of common features which can be used for the analysis. Assuming that we have a substrate of material A (e.g. GaAs) and a superlattice or MQW with a stack of AB layers, where B is the alloy (e.g. $Ga_{1-x}Al_xAs$), as illustrated in Figure 6.9, the rocking curve will show the following features:

1   A substrate peak from the A substrate.

2   A peak caused by the addition of Bragg reflections from the A and B components of the MQW. This is the zero-order or average mismatch peak, from which the average composition of the A + B layers may be obtained by differentiation of Bragg's law.

3   A set of subsidiary 'satellite' peaks symmetrically surrounding the zero-order peak, with spacing determined by the periodicity (total thickness of the repeating layers) of the MQW. These may also be regarded as the sum of the interference or gap fringes arising from interference between each of the layers comprising the MQW.

These peaks may be analysed to give much of the information required by the crystal grower. The methods of analysis have been developed by Segmuller et al.,[15] Kervarac et al.[16] and by Fewster[17,18] whose treatment we use here.

First we shall give a qualitative treatment of the effects of multilayers in general upon diffraction. As we have noted above, the term 'superlattice' may be taken literally – we have a large-spacing lattice with the periodicity of the spacing of the units making up the superlattice. If this is 200 Å, a typical value, we could use this as a 'Bragg' plane, apply the Bragg law and with CuK_α radiation there should be a first-order Bragg peak at $\theta_B = 0.221°$. Such peaks can indeed be seen with grazing-incidence reflectometry.

The number of units in an MQW will be much more limited than the number of atomic planes sampled by the X-ray beam in a standard reflection. The intensity will be low, but also the MQW will behave as a 'thin crystal' – the reciprocal lattice points will be extended into rods perpendicular to the crystal surface. This will broaden the reflection, and thus the width of each satellite peak is determined by the number of units in the MQW. It has even been possible, by careful analysis, to count the number of units more accurately than the prediction by the crystal growers.

The number of satellite peaks will depend on the shape of the interface between the units. It is convenient to think of the diffraction pattern in the kinematic approximation as the Fourier transform of the structure. If the layers in the units were graded so that the overall structure factor variation were sinusoidal, this would have only one Fourier component and thus only one pair of satellites. If the interface is abrupt, this is equivalent to the Fourier transform of a square wave, which consists of an infinite number of odd harmonics; the corresponding diffraction pattern is also an infinite number of odd satellites. The intensities of the satellites therefore contain information about the interface sharpness and grading.

If there are variations in the period (period dispersion) then there are in effect different periods within the superlattice. The zero-order peak for each of these must be the same – because it is simply the average mismatch of the alloy – but the distribution of intensity in the various interference fringes will be slightly different. This will tend to affect the higher-order satellite peaks more than the lower orders, and if measured with a low resolution instrument there will appear to be an increase in the width of the satellite peaks with order of the satellite, having taken out any instrumental functions.

6.6.1 *Average mismatch*

Subject to the caveat that there can be a significant shift in peak position when the total layer thickness is sub-micrometre we can determine the average composition of the MQW using the zeroth order, or average mismatch, peak. Asymmetric reflections are often used, both to determine any relaxation and to enhance the diffraction from thin layers. Let the period of the superlattice in real space be Λ, and the thickness of layers of $A_xB_{1-x}C$ of composition x_1 and x_2 be D_1 and D_2 respectively. Then

$$\Lambda = D_1 + D_2 \tag{6.3}$$

and

$$D_1x_1 + D_1x_2 = x_{av}\Lambda \tag{6.4}$$

where $x_1 + x_2 = 1$. Therefore

$$D_2 = \Lambda\left(x_{av} - x_1\right)/\left(1 - 2x_1\right) \tag{6.5a}$$

$$D_1 = \Lambda\left(x_{av} - x_2\right)/\left(1 - 2x_2\right) \tag{6.5b}$$

If the total A + B component of the MQW is less than about $0.25\,\mu m$, then the zero-order peak does not appear and dynamical-theory simulation must be used to determine the mean composition.

6.6.2 *Periodicity*

For X-ray wavelength λ, let L_i and L_j be two diffraction orders, (e.g. 5 and 7) and θ_i and θ_j the angles at which these orders diffract; then the following relationship holds:

$$\Lambda = \left(L_i - L_j\right)\lambda / \left[2\left(\sin\theta_i - \sin\theta_j\right)\right] \tag{6.6}$$

Λ then follows from the measurement of any two satellite peak positions, or, better, the measurement and averaging of several.

One may wish to know separately the average widths of each part of the unit (well and barrier); change in this ratio with no change in the overall period will not affect the satellite positions, but only their intensities. In effect, we have a large unit cell, the superlattice or MQW unit, and are varying the structure factor within this cell. The procedure is then to calculate the structure factors of all the observable satellites of the superlattice unit, using the normal formula to sum over atoms in a unit cell, and compare these with the integrated intensities actually measured after correcting for any instrumental effects. The calculation is iterated until the difference between the calculated and measured intensities is minimised as a function of the well to barrier ratio. The difference is expressed as the usual R factor

$$R = \sum\left\{\left|F_o\right| - \left|F_c\right|\right\}/\left|F_o\right| \tag{6.7}$$

where F_o and F_c are the observed (integrated intensities) and calculated structure factors.

The procedure is shown in detail in Fewster[18], who states that it is accurate to about 5% of a monolayer, and that it can be used to determine the incommensurability in well widths (i.e. widths that are not an integral number of lattice planes but formed from fractions of a monolayer).

It is worth noting that there are two particularly useful cases with sharp interfaces:

1   If the well to barrier ratio is 1:1 then even-numbered satellites are absent. This corresponds to the Fourier transform of a square wave, which contains only odd harmonics.

2   If the well to barrier ratio is 2:1 or 1:2 then third (sixth, ninth, ...) harmonics are absent.

6.6.3   *Thickness or period dispersion*

Fewster obtains this from the increase of satellite widths as a function of satellite order. The true integral breadths or FWHM values are first found; with a double or multiple crystal diffractometer this is simple measurement, and with a powder diffractometer the instrument function must first be found (by finding the zeroth-order integral breadth and subtracting).[19]

Equation (6.6) may be simplified to

$$\Lambda = \left(L_i - L_j\right)\lambda/2\!\left(\cos\theta\right)\!\left(\Delta\theta\right) \tag{6.8}$$

on the assumption that the satellite peaks are near the Bragg peak and that over that range $\cos\theta$ does not change appreciably. We may differentiate equation (6.8) to obtain

$$\Delta\Lambda = \left(L_i - L_j\right)\lambda d\!\left(\Delta\theta\right)\!\big/\!\left(\cos\theta\right)\!\left(\Delta\theta\right)^2 \tag{6.9}$$

The meaning of $d(\Delta\theta)$ is that it is the difference in the angle within the satellite – the satellite broadening – for a period difference of $\Delta\Lambda$. The FWHM of the Gaussian distribution of periods is given by

$$\sigma(\Lambda) = \left(L_i - L_j\right)\lambda b_s\big/\!\left(\cos\theta\right)\!\left(\Delta\theta\right)^2 \tag{6.10}$$

where b_s is the statistical average for all the measured satellites.

This approach does give plausible results but the physical mechanism is unconvincing, since the individual peak widths of both satellite and intermediate thickness fringes must be determined by the *total* thickness of the superlattice structure. Superlattice reflections can be thought of as a constructive interference between thickness fringes from successive units of the superlattice, as is easily seen by modelling a series of superlattices with increasing numbers of repeat units on a simulation program. Such modelling shows that the individual reflection widths change little with the introduction of period dispersion but that the relative intensities of different maxima in the thickness oscillations are very sensitive to irregularities in the period. Often, a small period dispersion transfers intensity from the superlattice peak to the immediately adjacent thickness fringes, giving the impression of peak width increase if the resolution is not sufficient. It is therefore by

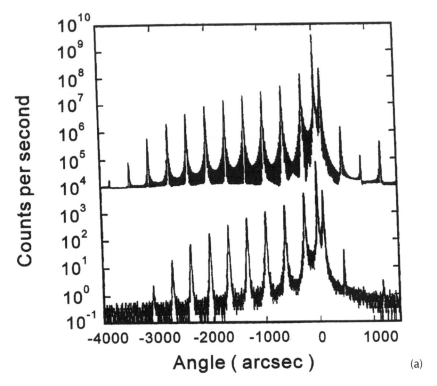

Figure 6.13 (a) Experimental and simulated curves for a Si–Ge 20-period superlattice, (b) Enlargement showing the thickness fringes between the −2 and +1 satellites. Upper curve is a RADS simulation of the superlattice structure with 2% period dispersion introduced, showing irregular changes in the amplitude of the thickness fringes but little change in width of satellite or subsidiary fringes. (Courtesy Dr A.R. Powell)

examination and simulation of the intensities of fringes in between the satellite reflections that we may determine period dispersion. Figure 6.13 shows such a determination, in the a Si–Ge superlattice, showing the details in between the −2 and +1 satellites. The simulated curve with 2% dispersion is not intended to be a perfect match, but to be clearly more irregular and showing a greater period dispersion than the experimental curve. Note the regularity of the fringes in the latter and the narrowness of the satellite reflections both in the experiment and in the simulation.

6.6.4 *Interface roughness and grading*

Roughness and grading cannot be distinguished in a double-axis rocking curve. Fewster's treatment is to assume that roughness is confined to less than one monolayer (over the coherence width of the beam) and to take up the rest of the combined parameter with grading. The introduction of sub-monolayer roughness permits the introduction of an incommensurate well width, found as described above. The roughness is introduced by adding coherently the structure factors of two commensurate periods, with well widths straddling the average

143

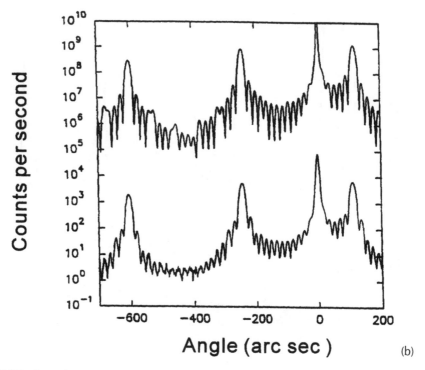

Figure 6.13 (*cont.*)

value. These are then scaled with the previously determined Gaussian dispersion of periods (equation (6.10)) and added to give the overall structure factor for that satellite. The structure factor of each satellite is computed and compared with experimental intensities, again using R factors, with the variable being the composition grading away from the interface, altered one layer at a time. Clearly this grading may be as complex as desired given ample computing time, but a linear grading over a few layers usually suffices. The effect of such grading can have a very significant effect on the relative heights of successive satellites. It can, for example, result in very substantial asymmetry in the positive and negative satellite peak intensities. An extreme example of this is shown in Figure 6.14. An example which has been quantified by means of simulation is shown in Figure 6.15 which gives a series of simulated rocking curves, due to Müller,[20] in which grading of the well walls is included. It is clear that such a change in the lattice parameter profile is necessary in order to obtain a good fit between simulated and experimental data.

An exemplary study of superlattice rocking curves was made by Gillespie *et al.*[21] on Si/GaAs. Analysis of the data revealed that an excellent fit could be obtained between simulation and experiment if it was assumed that the silicon layers were non-integer numbers of monolayers in thickness. This non-physical situation could be resolved by assuming that there existed a monolayer at the interfaces which had mixed occupancy of Si and Ga or As atoms. The phase shift introduced into the X-ray waves by such an alloy layer led to an equally good fit between simulation and experiment.

More recently Holý *et al.*[22] have developed a theory which treats roughness

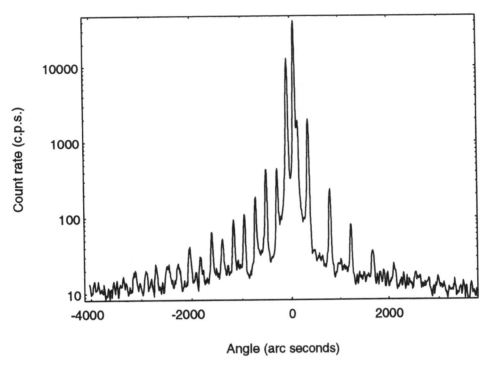

Figure 6.14 Rocking curve of a superlattice showing very substantial asymmetry between the $+n$ and $-n$ satellite peaks. Alternate $+n$ satellites are absent

properly and shows that true roughness results in broadening of the double-axis superlattice satellite peaks. The broadening varies as n^2, where n is the satellite order. Roughness does not therefore broaden the zero-order peak. However, the data produced by Holý *et al.* suggested that the broadening was constant with satellite order and that the broadening was due to a mosaic in the underlying crystal structure.

6.6.5 *Simulation of superlattice structures*

In simulating a superlattice structure, there is a need to establish initial input parameters for a first trial. The following procedure is not exclusive, but is that adopted by the authors.

1 Determine the superlattice period from the spacing of adjacent satellites.

2 Estimate the ratio of the well-to-barrier from the relative integrated intensities of successive satellites. Recall that the satellites correspond to a Fourier transform of the periodic structure. This done, we have input parameters for both well and barrier thickness.

3 Determine the average composition x_{av} from the position of the zero-order peak.

4 Use these parameters to estimate the individual compositions x_1 and x_2 of the superlattice using equations (6.5).

5 Input these parameters into a simulation program and begin iteration to a best fit.

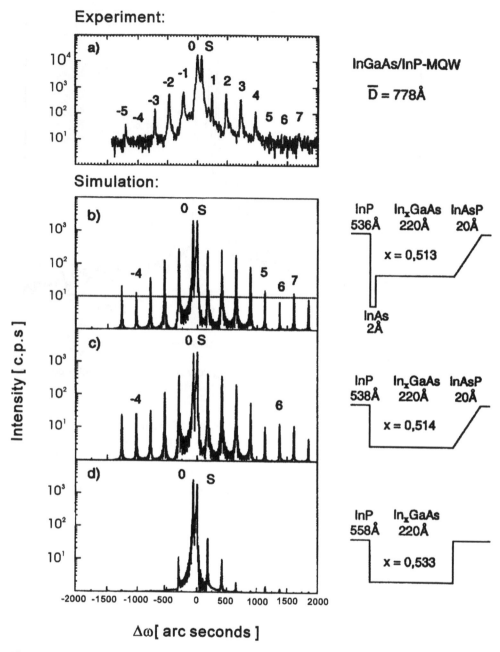

Figure 6.15 Experimental and simulated rocking curves with differing lattice parameter profiles through the well–barrier wall. (Courtesy R. Müller, University of Munich)

An example of the success of such fitting procedures is given in Figure 6.16, which shows experimental and simulated rocking curves of a $Cd_{1-x}Mn_xTe/CdTe$ superlattice grown by molecular beam epitaxy on InSb. Not only is the large-scale structure very well modelled but, subject to inclusion of a thin interface layer of In_2Te_3, there is an excellent agreement in the detailed fringe structure around the satellites.[23]

146

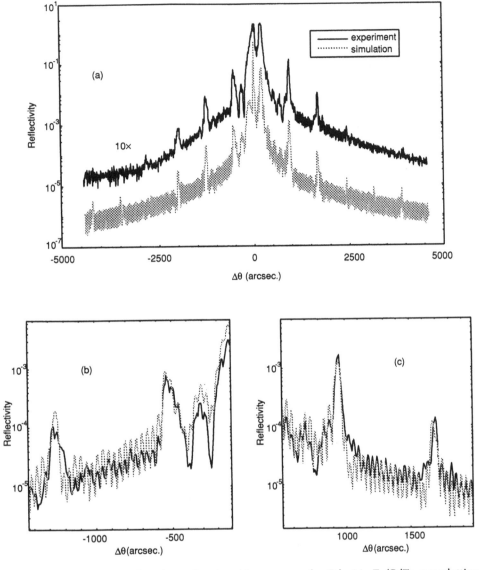

Figure 6.16 Experimental and simulated rocking curves of a $Cd_{1-x}Mn_xTe/CdTe$ superlattice grown on InSb The lower figures show enlargements of the region around the −1 and +1 satellites respectively.[23] (Courtesy Dr C. R. Li)

6.7 Summary

Thin epitaxial layers display a rich variety of X-ray optical phenomena which can be exploited for materials characterisation. Superlattice structures in particular display rocking curves which are sensitive to interface strains and these provide unique tools for investigating the structure–property relationships in an important class of electronic materials.

References

1. J. E. AYERS, J. Crystal Growth, **78**, 3724 (1995).
2. K. DUROSE & H. TATSUOKA, Inst. Phys. Conf. Ser., **134**, 581 (1993).
3. P. F. FEWSTER & C. J. CURLING, J. Appl. Phys., **62**, 4154 (1987).
4. C. R. WIE, J. Appl. Phys., **66**, 985 (1989).
5. D. K. BOWEN, N. LOXLEY, B. K. TANNER, L. COOKE & M. A. CAPANO, Mater. Res. Soc. Symp. Proc., **208**, 113 (1991).
6. M. J. HILL, B. K. TANNER, M. A. G. HALLIWELL & M. H. LYONS, J. Appl. Cryst., **18**, 446 (1985).
7. C. D. MOORE, PhD thesis, Durham University (1997).
8. M. H. LYONS, J. Crystal Growth, **96**, 339 (1989).
9. U. BONSE & M. HART, Appl. Phys. Lett., **6**, 155 (1965).
10. D. K. BOWEN, in: X-ray and neutron dynamical diffraction: theory and applications, eds. A. AUTHIER, S. LAGOMARSINO & B. K. TANNER (Plenum Press, New York, 1996) p. 381.
11. H. HOLLOWAY, J. Appl. Phys., **67**, 6229 (1990).
12. L. TAPFER & K. PLOOG, Phys. Rev. Phys. Rev. B, **40**, 9802 (1989); L. TAPFER, M. OSPELT & H. VON KANEL, J. Appl. Phys., **67**, 1298 (1990).
13. G. S. GREEN, B. K. TANNER & P. KIGHTLEY, Mater. Res. Soc. Symp. Proc., **208**, 315 (1991) G. S. GREEN, B. K. TANNER, S. J. BARNETT, M. EMENY, A. D. PITT, C. R. WHITEHOUSE & G. F. CLARK, Phil. Mag. Letts., **62**, 131 (1990).
14. C. R. LI, P. MÖCK, B. K. TANNER, D. ASHENFORD, J. H. C. HOGG & B. LUNN, Il Nuovo Cimento D **19**, 447 (1997).
15. A. SEGMULLER, P. KRISHNA & L. ESAKI, J. Appl. Cryst., **10**, 1 (1977).
16. J. KERVARAC, M. BAUDET, J. CAULET, P. AUVRAY, Y.Y. EMERY & A. REGRENY, J. Appl. Cryst., **17**, 196 (1984).
17. P. F. FEWSTER, Philips J. Research, **41**, 268 (1986).
18. P. F. FEWSTER, Semicond. Sci. Tech., **8**, 1915 (1993).
19. P. F. FEWSTER, Repts. Prog. Phys., **59**, 1339 (1996).
20. R. MÜLLER, University of Munich, private communication.
21. H. J. GILLESPIE, J. K. WADE, G. E. CROOK & R. J. MATYI, J. Appl. Phys., **73**, 95 (1992).
22. V. HOLÝ, J. KUBENA, E. ABRAMOF, A. PRESEK & E. KOPPENSTEINER, J. Phys D: Appl. Phys., **26**, A146 (1993).
23. C. R. LI, B. K. TANNER, D. ASHENFORD, C. H. J. HOGG & B. LUNN, J. Appl. Phys., **82**, 2281 (1997).

7

Triple-axis X-ray Diffractometry

In this chapter we show how restricting the angular acceptance of the detector adds another dimension to the information available from high resolution diffraction techniques, enabling strains and tilts in a sample to be identified separately. We explain the reciprocal space representation of triple-axis diffraction maps, and give several examples of its application to materials characterisation.

7.1 Introduction

A double-axis system, with or without a monochromator, uses an open detector and therefore integrates the scattering from the specimen over all angles within its aperture. Whilst this is relatively quick and convenient, it loses information; in particular, the scattering from bent or mosaic crystals occurs at different settings of the specimen crystal for a given d spacing, and details such as thickness fringes or narrow peaks can be lost or blurred. In some cases, such as Figure 7.1, the width of the slit before the detector determines the apparent material quality. This ambiguity can be removed by analysing the direction of the scattered X-rays from the crystal in the triple-axis geometry.

In the triple-axis mode, an analyser crystal is placed after the specimen and before the detector. This is mounted on an axis concentric with the specimen and is scanned independently of the sample. The experimenter can then map the intensity distribution with respect to the direction of the radiation scattered by the specimen. This not only removes the complication of a possibly bent or mosaic specimen, but also enables one to distinguish scattering from various sources. For example, scattering due to defects occurs in a different direction in space from scattering from the perfect crystal and from a map of the scattering as both the specimen and analyser are rotated, this can be measured quantitatively. Scattering from a rough surface can be separated from the perfect crystal scattering and most importantly, strain or mismatch may be distinguished from tilt or mosaic spread.

7.2 Instrumentation

Figure 7.2 shows the schematic arrangement for triple-axis diffraction. The three axes are those controlling the beam conditioner (and hence the input beam), the specimen, and the analyser, respectively. Although the term 'triple crystal' is sometimes used for this arrangement, but it is best avoided since it may erroneously be

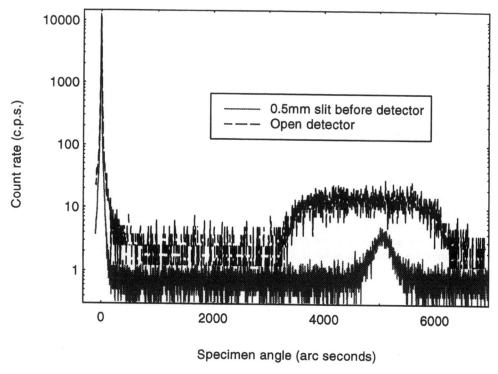

Figure 7.1 Double-axis rocking curves from a GaN epitaxial layer on (111) orientation GaAs. The measured rocking curve width is determined by the detector aperture

thought of as applying to multiple-crystal beam conditioners in an ordinary (i.e. double-axis) high resolution experiment. It is seen that the only difference between the triple- and double-axis instruments is the provision of a means to restrict the angular acceptance of the detector.

7.2.1 *Mechanical precision*

The angular precision required is substantial. The most important factor is for the detector axis to track the specimen axis accurately and continuously, to better than an arc second for most semiconductor work, corresponding to a reciprocal space resolution of $5 \times 10^{-6} \text{Å}^{-1}$. Random errors in the tracking result in 'noise' in reciprocal space, while systematic errors give rise to systematic distortions of the reciprocal space map. Artefacts due to backlash and eccentricity of gear trains are noticeable, and direct axis encoders are much preferred. Absolute accuracy is less important since reciprocal space maps are normally viewed, rather than measured, and errors around 1% are unlikely to be noticed.

7.2.2 *The beam conditioner*

The beam conditioner has two jobs: to condition the beam in angle and in wavelength. The angular conditioning may be improved by using channel-cut crystals[1]

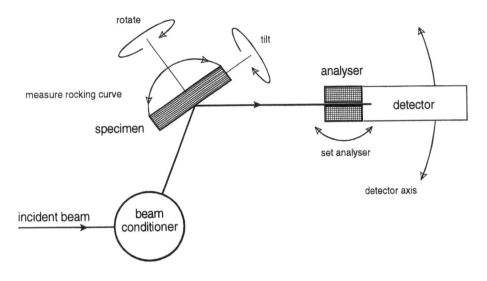

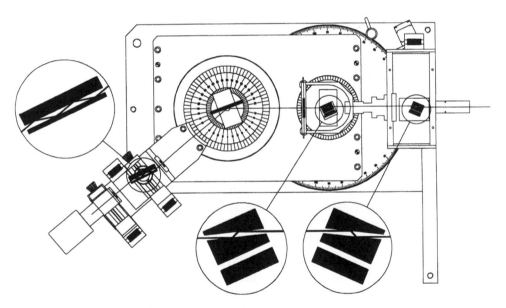

Figure 7.2 Diagram of a high resolution triple-axis instrument

thus reducing the streak in reciprocal space discussed in section 7.7. In practice, four reflections are required to make the streak scarcely noticeable. Variation of wavelength also affects the resolution, since all the important parameters scale with $1/\lambda$. Eliminating the $K_{\alpha 2}$ component is normally sufficient, which may be achieved with a slit after the conditioner, but any of the monochromating arrangements described in Chapter 2 may be used if the intensity is sufficient.

7.2.3 **The analyser and the detector**

The angular restriction may be performed in a number of ways: by a slit, Soller slit, single or channel crystal or by an area detector. The channel-cut analyser crystal gives two advantages if intensity is not at a premium. As with the beam conditioner crystal it reduces the instrumental streak in reciprocal space. It also preserves the direction of the beam scattered from the sample, simplifying the alignment. If only crude resolution is required, a slit may be sufficient (for example, to remove most of the effects due to the mosaic spread in the analysis of poor crystals). A single germanium crystal gives a high overall intensity.

For rapid low-resolution work, the use of an area detector both provides sufficient angular resolution and gives rapid data collection.[2,3] Otherwise, any detector may in principle be used, but a detector with a wide dynamic range and with good performance at the low-signal end is highly desirable, since effects such as diffuse scatter are weak and the X-ray optics necessarily reduce intensity.

7.3 **Setting up a triple-axis experiment**

At first sight, the complexity of yet another Bragg reflection seems appalling. However, this is not the case with modern instruments with computer-controlled alignment. With a little experience, triple-axis measurements will be no more challenging than most double-axis ones. Whilst the details of alignment will depend on the particular instrument, the principles are common for systems using crystal analysers:

1 Align the incident beam monochromator to deliver a beam passing accurately over the second (specimen) axis. This procedure is exactly as for high resolution double-axis diffraction.

2 With the specimen removed drive the detector axis (which carries the analyser) to the zero angle position. Then independently orient the analyser crystal to its approximate Bragg angle. Rotate the analyser crystal, independently of the detector axis, to locate the Bragg peak.

3 Use the tilt adjustment on the analyser to maximise the intensity in the Bragg peak. This must be done iteratively, tilting and relocating the Bragg peak. When the maximum has been reached, the analyser is correctly set with Bragg planes parallel to those of the beam conditioner and monochromator. Set the analyser onto the Bragg peak maximum.

4 Translate (or rotate) the analyser away from the detector so that the instrument is in double-axis mode. It is important that this motion should be reproducible so that the analyser may be returned within about an arc second.

5 Place the specimen on the second axis and with the detector axis rotated to twice the specimen Bragg angle, locate the Bragg peak of the specimen as in a double-axis experiment. Perform any necessary tilt adjustments of the specimen exactly as for double-axis diffraction. (If you wish you may perform a double-axis experiment at this stage.) Set everything onto the maximum of the Bragg peak.

6 Finally translate the analyser back in front of the detector and if necessary make a final small adjustment of the analyser rotation to maximise the signal once more. You are now set to perform one of several types of triple-axis experiment.

Triple-axis measurements normally take this 'everything at maximum' position as the 'zero' setting for reciprocal space maps. Two qualifications are needed. The position will of course not be the origin of reciprocal space but the reciprocal lattice point corresponding to the reflection used. Secondly, in the usual Bragg case it will be displaced by a few arc seconds to higher angles by the refractive index effect in dynamical scattering.

7.4 Separation of lattice tilts and strains

Triple-axis scattering enables the user to distinguish between tilts and dilations. This may be seen by considering the specimen shown in Figure 7.3, containing regions which are tilted with respect to each other, i.e. sub-grains and those that are strained or mismatched, e.g. ternary layers. By the Bragg law, the scattering angle defines the d spacing that is being examined. As the specimen is rotated, differently tilted regions will satisfy the condition for diffraction in sequence and the scattered intensity gives a measure of the distribution of tilts. Regions of the crystal where the Bragg plane spacing differs will never give rise to strong scattering when only the specimen is rotated.

Suppose we now perform a scattering experiment in which the specimen and analyser are scanned in synchronisation. Specifically, the analyser is scanned at

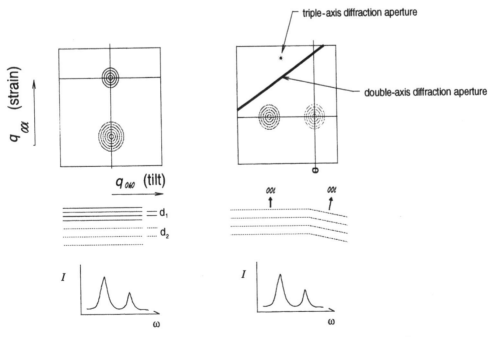

Figure 7.3 Triple axis measurements; real and reciprocal space representations

twice the rate of the specimen (a θ–2θ scan) both starting from zero. Suppose that a region of the specimen of lattice parameter d is set to diffract; then, as the analyser is set at twice this angle, intensity reaches the detector. If we now perform a θ–2θ coupled scan, any region of the specimen which also has lattice parameter d but is tilted with respect to the original region will never provide scattering which reaches the detector; the analyser setting will never be correct (unlike the example in the previous section). However, another region of the sample with lattice parameter d' may come to a position where the Bragg angle is satisfied. Now the analyser is at twice this angle and intensity reaches the detector. In this mode we record intensity from only parts of the crystal, but for heavily distorted materials such as gallium arsenide on silicon or small gap II–VI compounds, this provides a valuable measure of the range of lattice parameters present. In the case of ternary compounds such as cadmium mercury telluride, this provides a measure of the composition range, independent of the range of tilts.

Note also the resolution functions in reciprocal space of the double- and triple-axis measurements. In the double-axis rocking curves (shown) we cannot distinguish between the tilts and the dilations, but they are very different types of defect. The triple-axis resolution function in the dispersion plane is the intersection of two vectors – incident and diffracted beam – spread out by the angular width of the monochromator and analyser, respectively. A further blurring occurs in the plane normal to the figure, due to the vertical divergence, and the final component of the resolution function is the wavelength spread; this changes the lengths of the incident and diffracted beam vectors.

7.5   Measurement of kinematical scattering from a defective surface

The scattering from a distorted surface region is no longer governed by the dynamical scattering in the bulk of the crystal. In the bulk, due to multiple scattering processes, the refractive index is not quite unity and the result is that the position of the Bragg peak is shifted by a few arc seconds with respect to the value calculated by the Bragg equation. The kinematical scattering from the defective surface is, however, not shifted, nor does it have the same symmetry and in principle these effects can be distinguished.

In the double-axis geometry, the kinematical scattering lies within the envelope of that from the perfect bulk crystal and cannot be separated. However, if the specimen is set some way off the Bragg position, rotation of the analyser will pick up this scattering. The intensity is extremely sensitive to surface preparation. Figure 7.4 shows a broad diffuse peak due to polishing damage in gallium arsenide, superimposed on the sharp Bragg peak.

7.6   Measurement of distorted specimens

If a specimen is heavily dislocated it may be impossible to distinguish peaks in double-crystal rocking curves. Figure 7.5 shows a five-layer Si–Ge specimen grown with $0.5\,\mu$m thick layers, with the Ge content in each layer respectively 10, 20, 30, 40 and 50%; the aim being to produce a moderate amount of relaxation at each layer so that the top layer is fully relaxed 50% Ge but without excessive threading dislocations. This was achieved – the dislocation density was about

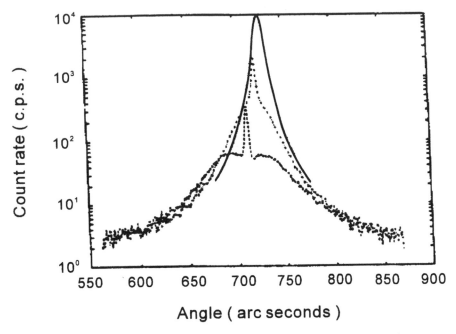

Figure 7.4 Defect peak in GaAs due to polishing damage. 004 reflection with CuK$_\alpha$. Triple axis θ–2θ scans with displacements of specimen from Bragg condition: solid (upper) line, 10″ displacement, next highest line, 20″ displacement, lowest line, 20″ displacement

$10^6\,\text{cm}^{-2}$ – but development of the process was not possible using only double-crystal diffraction even with a channel-cut collimator, as shown in the Figure 7.5(a). The range of the broad peak covers the 0–50% Ge contents, but no details are visible. In the triple-crystal setting, however, the separate layers are easily visible in a θ–2θ scan (Figure 7.5(b)). A single Ge crystal was used as the analyser for these weak signals.

Once such detail is available it is possible to scan in the perpendicular direction in reciprocal space, using a θ scan, to measure the tilt independently of the strain. This is shown in Figure 7.5(c), in which the analyser was set on the prominent left-hand peak in the triple-crystal curve. The tilt variation in this layer is seen to be about 1500″. Similar measurements on other peaks and correlation with crystal growth parameters can provide excellent feedback to the crystal grower.

The specimen-only rocking curve measures mosaic tilt very efficiently and it is intrinsically very narrow; it may be measured in no more time than the double-axis rocking curve and is about a quarter of its width. It may be used, for example, to characterise overall wafer quality. Figure 7.6 shows wafer maps of the width of the triple-axis specimen-only rocking curve as a function of position on the wafer. The distinction between the two classes of GaAs wafer is very easily seen.

For thick layers of highly mismatched systems, the double-axis rocking curves often turn out to be dominated by tilts. This is the case for the II–VI materials CdTe and Hg$_x$Mn$_{1-x}$Te grown on GaAs.[4] Another example is GaN on sapphire[5] shown in Figure 7.7. The tilt distribution can be modelled assuming that there is a random distribution of the edge components of the misfit dislocations normal to the surface. This results in a Gaussian distribution, which is quite well represented in Figure

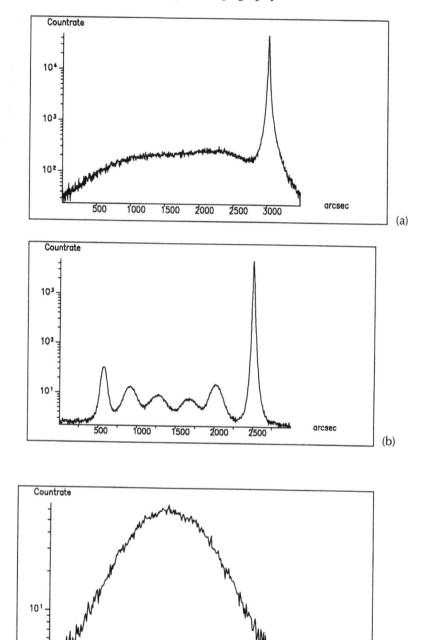

Figure 7.5 (a) Double-crystal rocking curve of the five-layer Si–Ge structure. (b) Triple-axis θ–2θ longitudinal scan of the same structure. (c) Specimen only (rocking curve) scan with the analyser set on the left-hand peak in the triple-axis curve. 004 reflection with CuK$_\alpha$ single-reflection Ge analyser for (b) and (c)

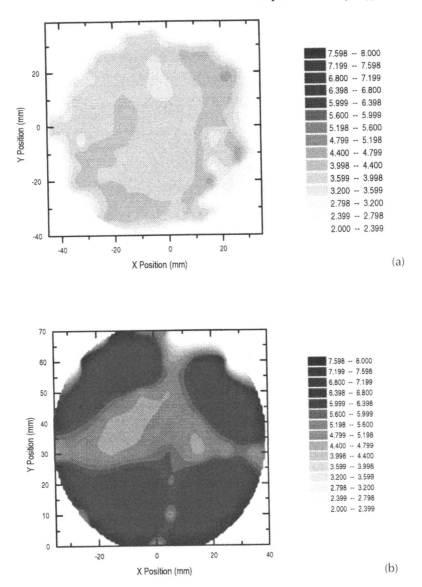

Figure 7.6 Wafer maps of the triple-axis rocking curve width (specimen-only scan) for GaAs grown by different processes. (a) A LEC GaAs wafer and (b) a VGZ GaAs wafer

7.7(a). Note that the coupled θ–2θ scan is very narrow in comparison with the specimen scan. This shows that there is little strain in the epitaxial layer but a large tilt distribution.

7.7 Full reciprocal space mapping

A full map of the scattering from the specimen can be made by recording intensity from a series of separate specimen and analyser positions, which are coupled so as

157

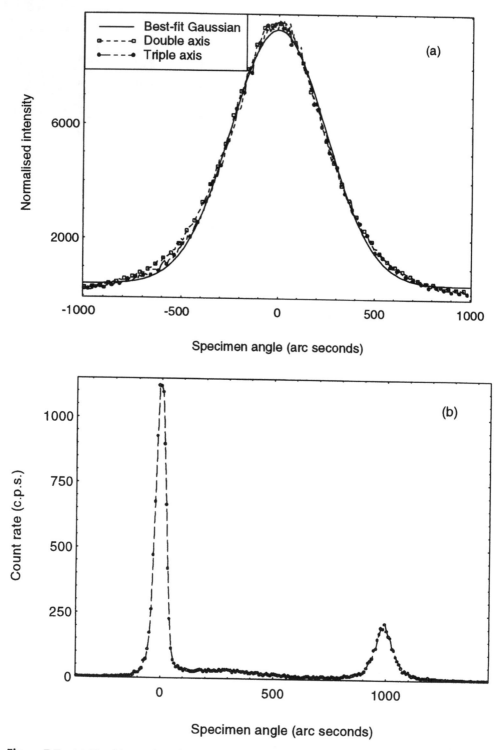

Figure 7.7 (a) Double- and triple-axis rocking curves (specimen scans) of a GaN epitaxial layer on (0001) oriented sapphire. 0002 reflection CuK$_\alpha$ radiation. (b) Coupled specimen-analyser scan in ratio 1:2. 022 (with 17.65° asymmetric cut) Si duMond beam conditioners, 111 symmetric Si analyser

to trace out a grid in reciprocal space. Such a contour map is shown schematically in Figure 7.8, after Iida and Kohra.[6] We may understand what is being measured in a triple-axis experiment with the aid of this figure. The angular positions of the incident beam and of the analyser define two vectors which define the scattering vector and hence angle. The angular position of the specimen defines the position of the diffracting planes whose scattering is being measured at this scattering angle. In Figure 7.8, the central point is the reciprocal space point of the diffracting planes, for example 004. The scattering is being measured in this instance from the small

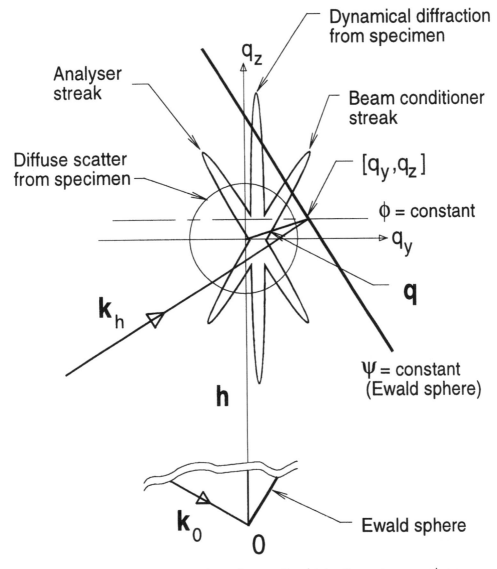

Figure 7.8 A scattering map in reciprocal space. Equal intensity contours are shown schematically, and the Ewald sphere is represented as a plane near reciprocal lattice points 0 and h. The dynamical diffraction from the specimen is displaced slightly from the relp and from the centre of the diffuse scatter by the refractive index effect

volume surrounding the point $[q_y, q_z]$. The scattering vector **Q** (not shown directly on Figure 7.8) may be considered as the sum of the 'ideal' scattering vector from the origin to point h, plus a deviation **q**; thus

$$\mathbf{Q} = \mathbf{h} + \mathbf{q} \tag{7.1}$$

The deviation vector **q**, with origin at the end of the reciprocal lattice vector **h**, has two components, q_y horizontal, positive rightwards going, and q_z vertical, positive upwards going. For the symmetric reflection, these components are related to the deviations of specimen ($\Delta\psi$) and analyser ($\Delta\phi$) from their zero positions at the nominal Bragg angle by the equations:

$$q_z = \Delta\phi \cos\theta_B / \lambda \tag{7.2}$$

$$q_y = (2\Delta\psi - \Delta\phi)\sin\theta_B / \lambda \tag{7.3}$$

Thus a scan of the specimen axis alone ($\Delta\psi$) affects only q_y and provides a scan from left to right in reciprocal space. A scan of the analyser affects both q_z and q_y and in fact sweeps along the Ewald sphere. A scan of q_z alone may be achieved by setting

$$(2\Delta\psi - \Delta\phi) = 0 \tag{7.4}$$

i.e. scanning the analyser at twice the rate of the specimen, usually called a θ–2θ scan (which notation has to be changed for the above description to avoid confusion between the axes).

Geometrically, we may visualise the above scans as follows. Scanning the specimen alone is equivalent to rotating the reciprocal lattice about its origin; the end of $\mathbf{k}_h$ thus describes an arc about 0, which at the scale of the drawing is a straight horizontal line. Scanning the detector alone is equivalent to changing the angle between $\mathbf{k}_0$ and $\mathbf{k}_h$, thus describing a scan along the Ewald sphere. The $2\Delta\psi - \Delta\phi$ scan moves vertically in reciprocal space arises because $\Delta\psi$ and $\Delta\phi$ are respectively the angles standing on the centre and the circumference of the Ewald sphere, subtended from the same arc, hence $2\Delta\psi = \Delta\phi$ by simple geometry of a circle.

The general case of the asymmetric reflection is a little more complicated and not usually given; the following derivation is due to Wormington.[7] The deviation of the scattering vector, **q**, from the reciprocal lattice point, **h**, can be calculated by considering the Ewald constructions shown in Figure 7.9. Figure 7.9(a) illustrates the case when the specimen angle, ψ, and the analyser angle, ϕ, are set so as to satisfy the Bragg condition, $2d \sin\theta_B = \lambda$. Figure 7.9(b) shows the Ewald construction after the specimen and analyser angles have been changed by $\Delta\psi$ and $\Delta\phi$ respectively. From equation (7.1) we have

On rearranging equation (7.1) we have

$$\mathbf{q} = \mathbf{h} - \mathbf{Q} \tag{7.5}$$

From the geometry of the Ewald construction shown in Figure 7.7(b) we can write $\Delta\mathbf{q}$ in terms of its Cartesian components (q_y, q_z);

$$q_y = 1/\lambda\{\cos(\phi + \Delta\phi - (\psi + \Delta\psi)) - \cos(\psi + \Delta\psi)\} - 1/\lambda\{\cos(\phi - \psi) - \cos(\psi)\} \tag{7.6}$$

and

$$q_z = 1/\lambda\{\sin(\psi + \Delta\psi) + \sin(\phi + \Delta\phi - (\psi + \Delta\psi))\} - 1/\lambda\{\sin(\psi) + \sin(\phi - \psi)\} \tag{7.7}$$

Expanding the trigonometric terms we may rewrite equations (7.6) and (7.7) as

$$q_y = 1/\lambda \left\{ \left[\cos(\varphi - \psi)\cos(\Delta\varphi - \Delta\psi) - \sin(\varphi - \psi)\sin(\Delta\varphi - \Delta\psi) \right] \right.$$
$$\left. - \left[\cos(\psi)\cos(\Delta\psi) - \sin(\psi)\sin(\Delta\psi) \right] \right\} - 1/\lambda \left\{ \cos(\varphi - \psi) - \cos(\psi) \right\} \tag{7.8}$$

$$q_z = 1/\lambda \left\{ \left[\sin(\psi)\cos(\Delta\psi) + \cos(\psi)\sin(\Delta\psi) \right] \right.$$
$$+ \left[\sin(\varphi - \psi)\cos(\Delta\varphi - \Delta\psi) + \cos(\varphi - \psi)\sin(\Delta\varphi - \Delta\psi) \right] \right\}$$
$$- 1/\lambda \left\{ \sin(\psi) + \sin(\varphi - \psi) \right\} \tag{7.9}$$

If $\Delta\psi$ and $\Delta\varphi$ are small, equations (7.8) and (7.9) can be simplified by using the small angle approximations, $\sin(\Delta\alpha) \approx \Delta\alpha$ and $\cos(\Delta\alpha) \approx 1$. We may finally write

$$q_y \approx 1/\lambda \left\{ \sin(\psi)\Delta\psi - \sin(\varphi - \psi)(\Delta\varphi - \Delta\psi) \right\} \tag{7.10}$$

$$q_z \approx 1/\lambda \left\{ \cos(\psi)\Delta\psi + \cos(\varphi - \psi)(\Delta\varphi - \Delta\psi) \right\} \tag{7.11}$$

where for symmetric reflection:

$$\psi = \theta_B, \quad \varphi - \psi = \theta_B$$

for asymmetric reflection with glancing incidence:

$$\psi = \theta_B - \phi, \quad \varphi - \psi = \theta_B + \phi$$

and for asymmetric reflection with glancing exit:

$$\psi = \theta_B + \phi, \quad \varphi - \psi = \theta_B - \phi$$

ϕ is the magnitude of the angle between the diffracting planes and the surface normal.

Several points arise from Figure 7.8. The main dynamical scattering from the specimen is a vertical streak, along the q_z direction, which in a good quality crystal has little extension in the q_y direction. This scattering is displaced from the reciprocal lattice point by the refractive index correction. However, diffuse scattering appears as a weak but broad, approximately annular scattering region centred on the reciprocal lattice point.

Streaks appear at $\pm\theta_B$ to the vertical axis. These are caused by the finite angular resolution of the beam conditioner and analyser crystals and would be absent if these crystals had rocking curves with no tails. They arise because at these settings of analyser and specimen, in addition to the true signal from the 'sampled region' of reciprocal space, a small number of the intense signals from the main peak pass through the tails of the analyser function. Hence the streaks point towards the main peak, and their direction is found by imagining a small oscillation of $\mathbf{k}_0$ about the origin and of $\mathbf{k}_h$ about the centre of the Ewald sphere. In a system with high resolution and very wide dynamic range, similar streaks at very low intensity levels can sometimes be seen from air scatter in the system e.g. Figure 7.11 – this has a similar effect in blurring the angular precision.[8]

Finally, we note that the standard double-crystal or high resolution rocking curve, in which we have no control over $\Delta\phi$ is in effect a horizontal scan through

161

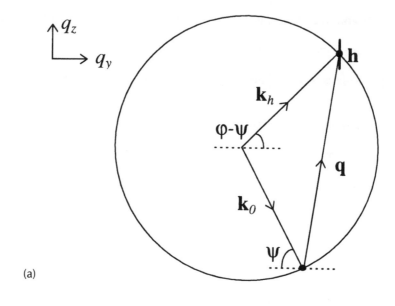

(a)

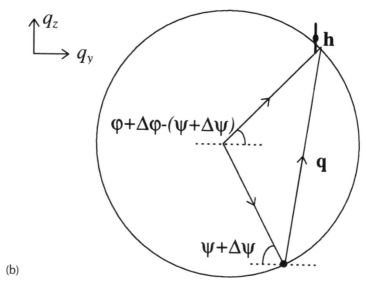

(b)

Figure 7.9 Ewald constructions, (a) at the Bragg condition, (b) off the Bragg condition

reciprocal space, integrating all intensities along the Ewald sphere. It is thus easy to see how the triple-axis instrument can obtain much more information.

7.8 Applications of reciprocal space mapping

We do not always need a synchrotron or a rotating anode X-ray generator to perform these studies. The resolution required determines the reflection used for the analyser. For low resolution measurements, a slit or a graphite crystal analyser can be used. An asymmetrically cut 022 Si or a 022 Ge analyser gives a modest resolution and for most systems this is a reasonable compromise between intensity and resolution in reciprocal space. A four-reflection symmetric Si 111 or 022 analyser provides a high resolution option for III–V or IV–IV semiconductor systems. The reciprocal space maps shown here were all taken in a few hours at most on a sealed-tube generator system.

A convenient data collection strategy is to perform a loop scan in which the specimen is set at a particular angle and then a θ–2θ scan is performed. The specimen position is then incremented and another θ–2θ or longitudinal scan is performed. In this way, one collects data in a square grid in reciprocal space. Appropriate software is then used to convert the data to reciprocal space coordinates using the equations given in the last section.

The first example we show is one of a defective region of a GaAs ingot grown in space under conditions of microgravity.[9] The main region of the ingot had good quality as indicated by 12″ wide rocking curves, but a second region, grown after a heater failure, exhibited broad, asymmetric rocking curves as shown in Figure 7.10(a). Quality is clearly low but the reason for this is not clear. The reciprocal space map (Figure 7.10(b)) shows that there are two areas of scatter. This feature is not revealed in the double-axis measurement! Within each region there is a large distribution of tilted regions, that is a large mosaic. The two regions appear displaced in the q_z direction, and are probably associated with lamella twins which are not quite coherent.

Growth of a thin epitaxial film of GaAs at low temperature results in the trapping of point defects which give rise to a larger lattice parameter than in the wafer. This comes about because excess As is incorporated, leading to an expansion of the basic GaAs lattice. The reciprocal space map shows that the epilayer has a differing lattice parameter (Figure 7.11) and that there is no tilt between epilayer and substrate. This can be seen from the fact that the 004 peaks of layer and substrate are vertically above each other in reciprocal space. There is equal broadening in the q_y direction in both substrate and epilayer peaks. This arises from long range curvature of the wafer due to the strain from the epilayer. (Note that the FWHM of the peaks for a longitudinal scan was 15 seconds, whereas the double crystal rocking curves were 60 seconds wide.)

Figure 7.11(b) shows the reciprocal space map after a high temperature anneal of the film. The effect of this has been to precipitate out the arsenic, resulting in the lattice parameter of the now stochiometric matrix reverting to that of the substrate. The scattering around the layer peak, which arises from the precipitates, is circularly symmetric and much more extensive than in the substrate.

In II–VI and III–V nitride compounds, epilayers are often rather imperfect. We saw in Figure 7.7 that the double-axis rocking curve with an open detector from a GaN epilayer on sapphire was almost identical to the triple-axis specimen only scan. This indicates that the strain distribution arises predominantly from tilts, not dilations and the θ–2θ scan is in contrast very narrow. A reciprocal space map (Figure 7.12) of the sample shown in Figure 7.7 reveals very clearly that the broadening of the double-crystal rocking curve comes predominantly from tilts, not dilations. The elongation of the scatter distribution in reciprocal space arises from the specimen-only scan being very much wider than the θ–2θ scan. Note that the layer peak does not lie exactly in the q_z direction with respect to the substrate peak. A simple θ–2θ

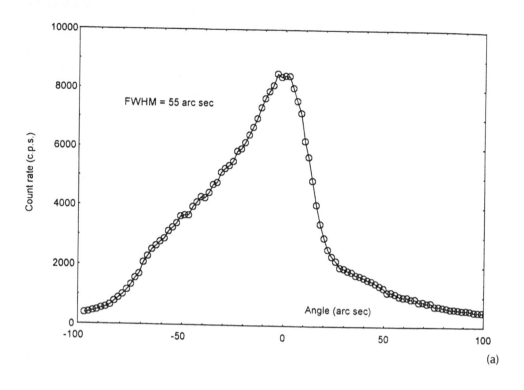

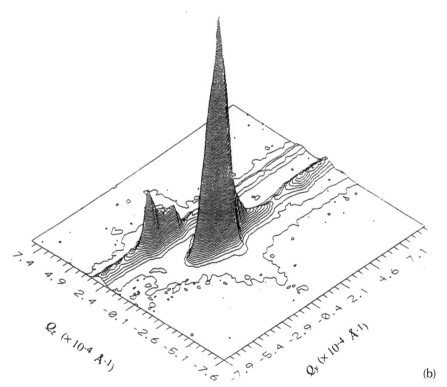

Figure 7.10 (a) Double-axis rocking curve of a microgravity-grown GaAs crystal after heater failure. (b) Equivalent triple-axis reciprocal space map CuK_α 004

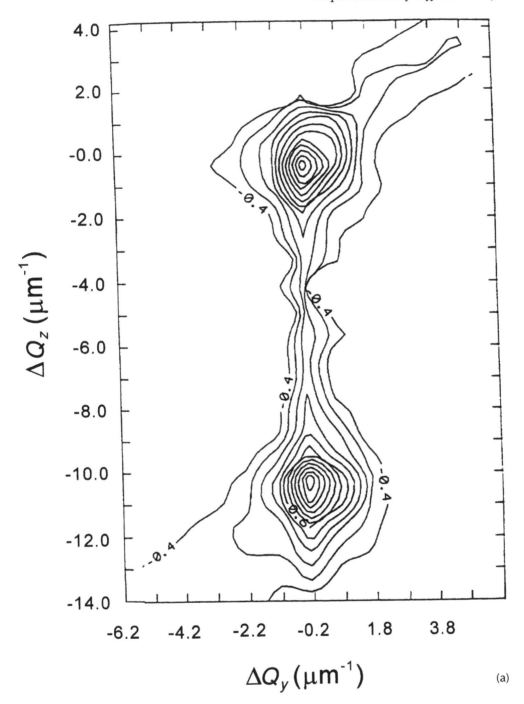

Figure 7.11 Reciprocal space map around 004 for a low temperature grown GaAs epilayer on GaAs. (a) As grown. (b) After high temperature anneal[10]

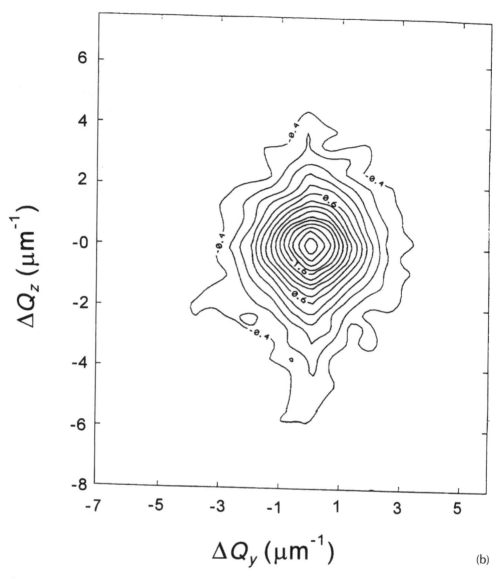

(b)

Figure 7.11   (*cont.*)

scan thus cuts through the shoulder of the layer peak, giving an incorrect impression of the height and width. Both the substrate and layer show a mosaic structure, although the latter distribution is wider than the former.

Figure 7.13 shows a reciprocal space map of a typical HEMT structure which contains a ~10 nm thick InGaAs layer on a GaAs substrate, together with a GaAs cap. We note that the thickness of the fringes associated with a thin layer now become contours of intensity, the maxima of which run along a line normal to the specimen surface. It can be seen immediately from Figure 7.9 that the substrate surface was offcut from the [001] direction, since the line of interference maxima does not coincide with the q_z direction. No significant broadening occurs of the interference fringes in this direction as the deviation in q_z increases. However, there

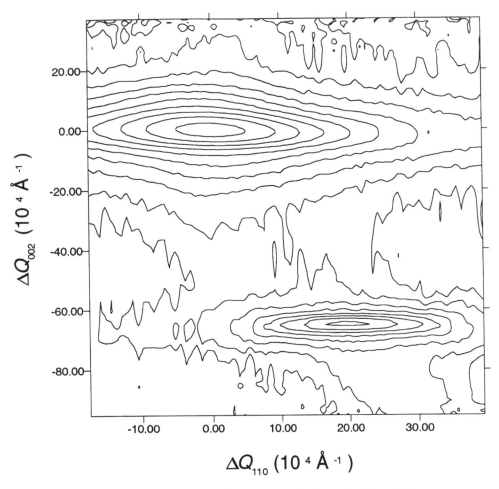

Figure 7.12   Reciprocal space map of the GaN sample shown in Figure 7.7

is a large region of strong diffuse scatter centred around the substrate reciprocal lattice point which is much stronger than normally found from substrate wafers. We would therefore ascribe this scatter to defects within the capping layer.

The final example (Figures 7.14 and 7.15) shows the complexity of structural detail that can be resolved through the use of reciprocal space mapping, and is the work of Goorsky's group (Eldredge *et al.*)[11] at UCLA. The specimen was a GaAs substrate, followed by a buffer layer in which the In content of InGaAs rose from 0 to 25% over just under two micrometres, in order to make a substrate with a new lattice parameter, on which a superlattice with (+,−) strain was then grown. This is a similar application to the five-layer Si–Ge structure discussed in section 7.6, but grown with continuous grading instead of steps. The object of the characterisation was to discover the degree of strain and relaxation in the buffer layer to see if this end was achieved. A further complication arises in that the substrate was tilted by 2° towards an <011> direction from the exact (001) orientation.

The method used by the group was first to determine the direction of substrate tilt by means of reciprocal space maps in each of the <110> directions contained in the

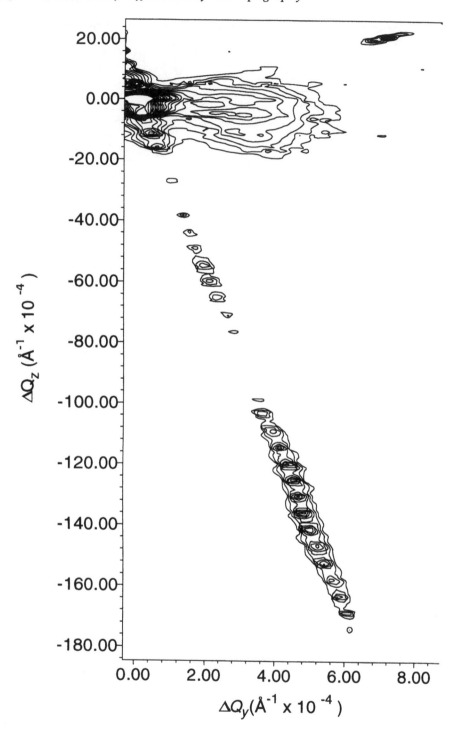

Figure 7.13 Reciprocal space map of a HEMT structure which contains an approximately 10 nm thick InGaAs quantum well and is capped by a thick GaAs layer. The substrate is (001) GaAs

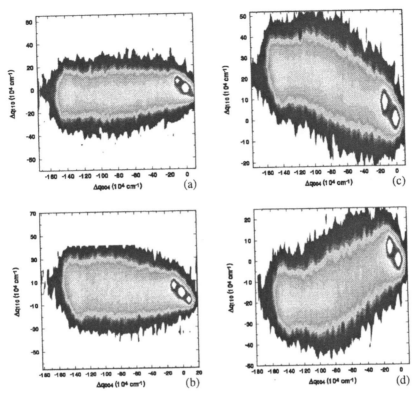

Figure 7.14 Reciprocal space maps of GaAs with graded InGaAs buffer layer around the 004 point with different directions of the incident beam. (a) Along [110], (b) along [1̄1̄0], (c) along [1̄10], (d) along [11̄0]. In this and the next figure the orientation is such that the origin is on the left (not the bottom) of the maps, hence the horizontal direction represents strain and the vertical direction, tilt

plane of the wafer. These are shown in Figure 7.14. A narrow slit was used instead of an analyser crystal, so some analyser streaks are seen near the substrate peaks (on the right of each map – the origin in this case is on the left-hand sides). It is seen that in (a) and (b) the beam is oriented perpendicular to the substrate tilt axis, as these maps show only strain. In (c) and (d) the effect of the grading can be seen, since both the tilt and the strain are changing, but these views are insufficient to make a complete analysis. For this we need an asymmetric reflection such as 224. These are shown for 110 and 1̄10 reflections in Figure 7.15, with the 004 maps as insets, and this time the superlattice peaks are also seen.

The interpretation is as follows. The diagonal streak in the 22̄4 reflection is not very helpful since it simply shows us that there is both tilt and strain present in the buffer layer, and we do now know whether the tilt was caused by the offcut or by the growth. However, the 2̄24 map was taken in the direction insensitive to tilt arising from the substrate offset and we can be sure of the interpretation. The region of intensity that runs radially from the substrate peak towards the origin must have a

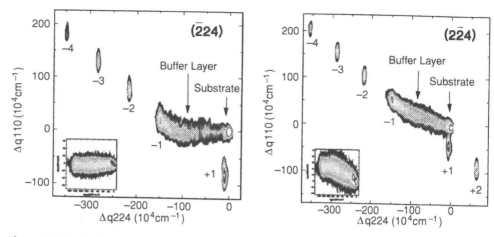

Figure 7.15   Reciprocal space maps of GaAs with graded InGaAs buffer layer around the 224 reciprocal lattice point. (a) [110] direction, (b) [$\bar{1}$10] direction. The 004 maps in those directions are shown in the insets

cubic structure of varying lattice parameter, i.e. it is fully relaxed. The part that runs diagonally and merges with the superlattice reflections must represent a tetragonally strained region, i.e. it is not relaxed. The fact that this strained region merges with the superlattice reflections means that the superlattice is modulating the strained region of buffer layer, not the relaxed region. Since the superlattice can only have coherency with the surface on which it is grown, not that further down, then the conclusion must be that the buffer layer is first relaxed (maybe due to too high a composition gradient), up to 20% In, and then becomes strained and coherent. The growth objective of a fully strained buffer layer was thus not achieved, but clear pointers to the process problems were discovered by this careful measurement.

7.9   Summary

In this chapter we have seen the new dimension that triple-axis diffractometry and reciprocal space mapping bring to the characterisation of complex materials. On the one hand they provide new ways of looking at materials that are not of good enough quality to use for the double-axis methods, and, on the other, they show how very complicated structural problems can be disentangled by this extra dimension.

References

1. P. ZAUMSEIL & U. WINTER, Phys. Stat. Sol. (a), **70**, 497 (1982).
2. L. R. THOMPSON, G. J. COLLINS, B. L. DOYLE & J. A. KNAPP, J. Appl. Phys., **70**, 4760 (1991).
3. S. R. LEE, B. L. DOYLE, T. J. DRUMMOND, J. W. MEDERNACH & R. P. SCHNEIDER, JR, Advances in X-ray Analysis, **38**, 201 (1995).

4. T. D. HALLAM, S. K. HALDER, J. M. HUDSON, C. R. LI, M. FUNAKI, J. E. LEWIS, A. W. BRINKMAN & B. K. TANNER, J. PHYS. D: Appl. Phys., **26**, A161 (1993).
5. T. LAFFORD, N. LOXLEY & B. K. TANNER, Mater. Res. Soc. Symp. Proc. **449**, 483 (1997).
6. A. IIDA & K. KOHRA, Phys. Stat. Sol. (a), **51**, 533–42 (1979).
7. M. WORMINGTON, Bede Scientific Instruments, private communication.
8. R. MATYI, Rev. Sci. Instrum., **63**, 5591 (1992).
9. N. LOXLEY, C. D. MOORE, M. SAFA, B. K. TANNER, G. F. CLARK, F. M., HERMANN & G. MUELLER, Adv. X-ray Analysis, **38**, 195 (1995)
10. J. M. HUDSON, PhD thesis, Durham University (1993).
11. J. W. ELDREDGE, K. M. MATNEY, M. S. GOORSKY, H. C. CHUI & J. S. HARRIS, JR, J. Vac. Sci. Technol., **B13**, 689 (1995).

8

Single-crystal X-ray Topography

In this chapter, we review the fundamental contrast mechanisms in X-ray topographs and describe the Berg–Barrett, section and projection topography techniques. Details are given of procedures for setting up and recording topographs. We discuss limitations on real-time topography and detector requirements. In an extensive section on the contrast of planar and line defects under various diffraction conditions, we provide some additional theoretical tools to aid interpretation.

8.1 Introduction

We have so far been concerned only with the measurement of the integrated intensity arriving at the detector as a function of angular setting of various crystals with no consideration of the spatial distribution of that intensity. Indeed, it was not until 20 years after the discovery of X-ray diffraction that the first papers appeared which reported studies of the variation of the diffracted intensity as a function of position across a crystal. In such experiments, the intensity variation across the diffracted beam is recorded, usually photographically, and thus a map of the scattering power is recorded as a function of position. Such X-ray diffraction topographs are analogous to transmission electron micrographs. Despite the name, the technique is *not* principally sensitive to *surface* topography; it is the topography of the crystal lattice planes that is examined.

As is the case in many branches of science, it is difficult to identify exactly when or by whom X-ray topography was invented. The early methods of Schulz[1] and Guinier and Tennevin[2] which recorded the intensity distribution across Laue spots were not capable of resolving individual dislocations. Normal to the incidence plane (i.e. that containing incident and diffracted beams) the spatial resolution is determined very simply by the projected size of the X-ray source. Owing to the large size of the X-ray sources and the need to place the film a significant distance from the specimen in order to separate the different Laue spots, only large-scale strains, such as are associated with scratches due to polishing damage, were resolved. However, as we will see in Chapter 10, with the building of synchrotron radiation sources it has now become possible to use these simple techniques for very fast, high resolution work.

Individual dislocation images were first observed in X-ray topographs by Lang[3] using his *section topography* technique and almost simultaneously by Bonse and Kappler[4] using the *double-crystal* method. The latter technique will be discussed in detail in Chapter 9 and in this chapter we will concentrate on single-crystal techniques which use the characteristic radiation from an X-ray tube and which are capable of high resolution imaging.

8.2 Contrast mechanisms

The aim of all X-ray topographic methods is to provide a picture of the distribution of the defects in a crystal, and the X-ray images may be thought of as arising in two ways. **Orientation contrast** arises when a region of the crystal is misoriented by an amount larger than the beam divergence (Figure 8.1). Then for the characteristic line, no diffracted intensity is recorded for region A when the Bragg condition is satisfied for the rest of the crystal. Thus there is an undarkened patch on the film. The angle of misorientation, projected into the incidence plane, can be determined by the angle that the specimen must be rotated in order to obtain strong intensity from the region A. This is very conveniently achieved in real-time using an electronic imaging detector but is tedious to perform with photographic recording. Orientation contrast may arise from the presence of twins, sub-grains and electric and magnetic domains and can be interpreted in a simple geometric manner without need of detailed dynamical diffraction theory calculations. Most of the contrast in double-crystal topographs is of this kind, due to the very low angular divergence of the beam at the specimen (see Chapter 9).

Another type of orientation contrast, discussed in Chapter 10, arises when continuous radiation is being used. It is most common at synchrotron sources but can sometimes be seen in single crystal topographs in the laboratory. Here the two regions A and B satisfy the Bragg condition for different wavelengths. The beams then make different directions in space and can either overlap or diverge, depending on the relative misorientation of regions A and B. Simple images occur where the

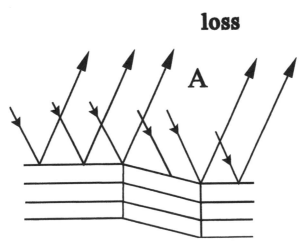

Figure 8.1 Orientation contrast from a monochromatic, collimated X-ray beam. Diffraction occurs only in region A for a particular angular setting of the specimen

crystal contains discrete mosaic blocks, but when the lattice distortion is continuous, the resulting contrast can be very complex and difficult to interpret.

The second major contrast mechanism is **extinction contrast**. Here the distortion of the lattice around a defect gives rise to a different scattering power from that of the surrounding matrix. In all cases, it arises from a breakdown or change of the dynamical diffraction in the perfect crystal. In classical structure analysis, the name *extinction* was used to describe the observation that the integrated intensity was less than that predicted by the kinematical theory. Around the defect, enhanced scattering was observed and this 'loss of extinction' is the origin of the name. The exact nature of the images can only be explained using dynamical diffraction theory and we will return to this in a later section.

8.3 The Berg–Barrett technique

This Bragg (reflection) orientation technique, developed in 1945 by Barrett[5] following earlier work by Berg,[6] is the simplest high resolution technique to set up in the laboratory. It uses an extended X-ray source and the specimen is adjusted so that diffraction conditions are satisfied for the characteristic K_α lines from a specific set of Bragg planes. Using crystals cut in such a way that the diffracting planes are aligned so that the incident beam makes a very small angle to the specimen surface (Figure 8.2), it is possible to place the photographic film within about 1 mm of the specimen surface. Best results are obtained for planes where the Bragg angle for the characteristic K_α lines is nearly 45°, so that the diffracted beam emerges almost normal to the specimen surface. This configuration has the convenient feature that it is very insensitive to orientation contrast. It also has two important consequences for spatial resolution.

The characteristic K_α line consists of a closely spaced doublet with a 2:1 intensity ratio between the $K_{\alpha1}$ and $K_{\alpha2}$ lines. This not a problem if a low spatial resolution recording medium, such as a fluorescent screen and image intensifier, is used but is a problem if high resolution nuclear emulsion plates are used. Then individual dislocation images are doubled because the diffracted beams from the two lines make different directions in space, giving rise to a type of orientation contrast. However, if the specimen to film distance is kept small, the divergence of the beams is negligible and sharp individual images can be recorded (Figure 8.3).

The second issue concerns the spatial resolution achievable with this setting. To a good approximation, the spatial resolution normal to the incidence plane δ is related to the specimen to source distance D, the specimen to plate distance L and projected height of the source H by the simple geometrical relation:

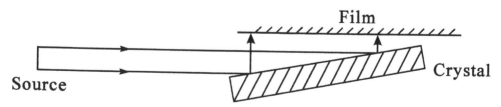

Figure 8.2   Schematic diagram of the Berg–Barrett technique

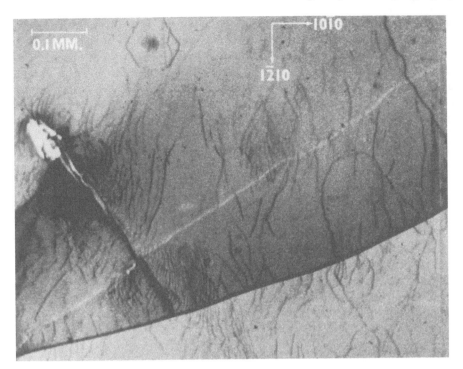

Figure 8.3 Berg–Barrett topograph showing individual dislocation images in a single crystal of zinc. (Courtesy B. Roessler)

$$\delta = HL/D.$$

This is illustrated in Figure 8.4. The benefit of being able to place the plate very close to the specimen is immediately clear. For example, with a projected source height 0.04 mm, distant 160 mm from the specimen, a spatial resolution of 1.25 μm can be achieved for a 5 mm specimen to plate distance. Despite its age and simplicity, the Berg–Barrett technique is still in use today, particularly for the initial assessment of crystals of new materials where the perfection is relatively low.

If Bragg planes cannot be chosen, or the crystal cut, so that the diffracted beam emerges almost normally to the specimen surface, the diffracted beam will pass through the recording photographic film at an angle. The consequent loss of resolution can be minimised by use of a very thin emulsion (e.g. 10 μm thick), but then a soft radiation such as CrK$_\alpha$ should be used to ensure that a significant amount of the

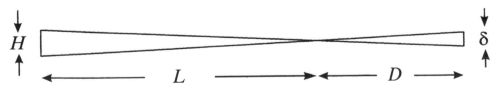

Figure 8.4 Schematic diagram showing the geometrical resolution limit set by the projected source height normal to the incidence plane

175

beam is stopped by the film. Use of soft radiation does often turn out to be convenient, as the large Bragg angles then encountered make it easier to find low-order, grazing incidence, asymmetric reflections with near 90° scattering angle.

An important feature of the use of soft radiation is that both the extinction distance and the absorption distance are small, and therefore the X-rays penetrate only a very small distance into the crystal. We therefore examine only a small slice of the crystal close to the surface. The strain fields of dislocations deeper into the crystal do not contribute significantly to the image and, in a transmission experiment, overlapping of the images leads to an upper limit of about $10^4 \, \text{cm}^{-2}$ on the dislocation density for individual defect imaging. With the Berg–Barrett technique this can be pushed to about $10^6 \, \text{cm}^{-2}$.

The large rocking curve width associated with the extended source makes it easy to set up the Berg–Barrett topograph. This can be done with a counter but is most easily achieved with a phosphor screen and TV camera. The strength of the beams is such that only modest gain is needed and the equipment can therefore be of quite low cost. Early workers used to look, under black-out conditions, for the image on a piece of fluorescent paper attached to a hand-held stick. Such practice is now illegal in most countries owing to the significant radiation dose received during this method of alignment.

8.4 Lang topography

8.4.1 *Section topography*

This transmission method[3] provides an image of a section through the crystal and as such enables the experimenter to study the three-dimensional distribution of defects. The beam from the spot of a fine-focus or microfocus source is collimated into a ribbon beam of width approximately $10 \, \mu\text{m}$ before the single-crystal specimen. This provides an incident beam of width small compared with that of the base of the Borrmann fan formed by extremes of the diffracted and transmitted beams with the crystal surface. [In other words, the beam width must be much less than $t \sin 2\theta_B$ where t is the specimen thickness and θ_B the Bragg angle.] The specimen is adjusted until a strong diffracted beam from the characteristic $K_{\alpha 1}$ line is obtained for the diffraction planes chosen and the photographic plate placed behind the specimen. Owing to the high collimation, simultaneous diffraction from the $K_{\alpha 1}$ and $K_{\alpha 2}$ lines cannot be achieved and there is thus no problem of image doubling. A diffracted beam slit prevents the main beam from striking the photographic plate. The geometric arrangement is such that the plate can be placed within about $10 \, \text{mm}$ of the specimen. For a source to specimen distance of $1 \, \text{m}$ and projected source height of $0.3 \, \text{mm}$ the geometrical resolution is then $3 \, \mu\text{m}$. Lang used a microfocus source to improve this to $1 \, \mu\text{m}$.

8.4.2 *Projection topography*

The section topograph provides an image from only a narrow strip down the crystal and in order to obtain an image of the whole crystal and still retain the high spatial resolution, Lang devised a goniometer in which the crystal and film were translated

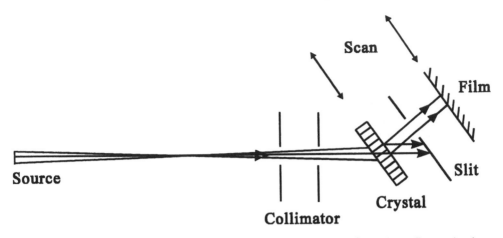

Figure 8.5 Schematic diagram of the projection topography configurations. For projection topographs, the specimen and film are translated across the beam. (Note that very narrow (~10 μm) slits are used for section topography whereas those for projection topography are typically 0.25 mm)

together across the beam[7] (Figure 8.5). Lang cameras, which are two-circle goniometers equipped with a precision translation stage and adjustable (or inter-changeable) incident beam slits, are available commercially and the technique has become the most widely used laboratory technique of X-ray topography. Section and projection topography can be performed with the same apparatus, although it is advisable to replace the very narrow slits necessary for section topography with wider ones when projection topographs are taken. This reduces exposure times very considerably.

8.5 Experimental procedures for taking Lang topographs

In experimental science there is no substitute for experience, but if certain pro-cedures are followed and adapted, a number of pitfalls can be avoided. The follow-ing sections are written for the benefit of beginners in topography, as this is a field relatively strange to practitioners of diffraction. While most of the material here can already be found in the literature[8,9] these works are out of print and much exists now only in the folklore.

8.5.1 *Setting up the crystal*

An important feature, often overlooked, is the technique of mounting the crystal. X-ray topographic techniques are necessarily sensitive to strain, and clumsy mounting will lead to difficulties. When the crystal is small, it is useful to mount it inside a suitable metal ring which is fixed to the goniometer. This enables a large number of reflections to be taken simply by rotating the mount. In plastic crystals many dislocations can be introduced and in brittle materials a radius of curvature as great as 2 m will prevent Bragg reflection from occurring simultaneously across a 1 cm wide sample. For many purposes a soft wax is a suitable mounting material provided that only the minimum necessary to hold the crystal is used. (Cenco Soft-Seal wax, melted by gently heating with a small soldering iron run cool works very well.) Fixing at one point only is essential for all cements. Quick-drying glues which contract on setting should be avoided. Epoxy resin or strain gauge cement can be successfully used for permanent mounting as they contract very little on setting, and a general rule is to apply the adhesive to the sample as far away from the region under examination as possible. At low temperatures a small quantity of varnish works well. Cyano-acrylate glues (super-glues), used with brittle materials such as silicon, can be removed through thermal shock by immersion in boiling water or liquid nitrogen.

8.5.2   *Setting the diffraction vector in the horizontal plane*

This is equivalent to the tilt adjustment in the high resolution diffraction experiment. It is, however, much less critical and can often be done by eye, for example with crystals conveniently displaying facets or edges oriented in definite crystallographic directions. (A draughtsman's variable angle set square is a useful tool.) For Lang topography this alignment to about 0.5° will be quite adequate, though fine adjustment can be made once the Bragg reflection has been found. Where the crystal has not crystallographically oriented edges, the whole goniometer should be transferred to a back-reflection Laue camera and the crystallographic orientation determined. Suitable rotation will then bring the required diffracting planes into a vertical orientation.

8.5.3   *Finding the Bragg reflection*

When initially aligning the camera to take off the X-ray beam at about 4.5° to 6° to the source, use of low X-ray power is advised and an absorber should be used in the beam. Most scintillation and proportional counter detectors saturate well below 10^6 cps. It is thus easy to believe that there are no X-rays present when, in practice, one is aligned on the unconditioned beam. During operation, high voltage – low current conditions give a good characteristic line intensity, and care should be taken to ensure that the critical voltage is exceeded. (A Bragg reflection will not be found with AgK_α at a tube voltage of 20 kV.)

Care should be taken to centre the crystal over the camera axis, but in any case it is advisable to reduce the distance between specimen and scintillator for initial setting. With a non-centred specimen, the effective 2θ angle of the counter will not be given by $2\theta_B$ read on the counter scale. A wide angular aperture avoids trouble but increases the background. Once the reflection is found, the angular acceptance should be decreased and the scintillator moved further from the specimen. This eliminates the possibility of having found a stray reflection from non-vertical diffracting planes. Even with computer control and automatic searching, 'some days it is easy, some days it is not'.[8]

Two peaks should be visible, the $K_{\alpha2}$ being half of the intensity of the $K_{\alpha1}$ peak. If this intensity ratio is not 2:1 it implies that the Bragg planes are not vertical and the orientation of the crystal in the plane normal to the incident beam should be adjusted.

When anomalous transmission studies are made of materials giving appreciable fluorescence it is sometimes difficult to detect a Bragg reflection from the thick crystal against the high background. In this situation the very intense surface reflection from the edge of the crystal can be used to great advantage. The crystal should be adjusted until it just cuts into the X-ray beam. Once the edge reflection is found, the crystal can be traversed into the thick region and the low visibility Bragg peak there will be seen. In the event of no peak being observed one can conclude fairly certainly that anomalous transmission is not occurring and the crystal is imperfect.

When the reflection has been found, the diffracted beam slits may be inserted. With low fluorescence it may be satisfactory to use only a straight edge to eliminate the direct beam, but in many cases the use of a slit is required. A broad rocking curve in which it is impossible to distinguish the $K_{\alpha1}$ and $K_{\alpha2}$ components is a good indication that the crystal is highly strained and the resulting topographs will not be works of art. They may, however, still be scientifically informative. Similarly, a rapid change in intensity on traversing, giving a strong signal over only a short range, is often indicative of a bent crystal. In the last circumstance, provided the curvature is not too great and hard radiation is used, it is possible to obtain good topographs by continuously adjusting the angular setting of the crystal to keep it exactly on the Bragg reflection. This is easy to implement with a simple algorithm in a computer-controlled camera.

8.5.4 *Recording the topograph*

As will become apparent, it is important to place the photographic plate as close to the specimen as possible. With rotating anode generators, care should be taken not to allow the full power of the beam to fall on the plate when stationary as this leads to an unsightly overexposed vertical line on the topograph. The presence of a horizontal stripe on the recorded topograph is often due to the presence of a second reciprocal lattice point lying on the Ewald sphere. It can be removed by a small rotation of the crystal about the diffraction vector as if to take a stereo pair.

8.6 Topographic resolution

As the criteria differ in the horizontal and vertical directions, they will be considered separately.

8.6.1 *Photographic resolution*

Unlike electron microscopy, the quality of X-ray mirrors is not sufficiently good to magnify an X-ray topographic image. The topograph must be recorded at a magnification of unity and subsequently enlarged optically. As dislocation images are upwards of a micrometre in width it is essential that a recording medium of comparable resolution is used. It is widely accepted that the Ilford L4 Nuclear Emulsions (undeveloped grain size $0.14\,\mu\text{m}$) have excellent characteristics for the recording of

X-ray topographs. Dental or standard X-ray film is fast and can be used to obtain a large area image, but the spatial resolution is poor. The Agfa D2 and D7 films, with grain size 2 and 7 μm, are often a good compromise between expense, speed and resolution. For topographs of large areas of semiconductor wafers they are very good. For high resolution optical images, a very thin emulsion would be chosen to obtain high resolution, but owing to the relatively poor absorption of X-rays there is a limit to the minimum thickness tolerable. For harder radiations, e.g. MoK$_\alpha$ and AgK$_\alpha$, 50 μm thick emulsion is used while for softer radiation 25 μm thickness is satisfactory. These constitute a compromise between efficiency and resolution. To obtain a resolution of 1 μm, the width of the projection of the ray in the emulsion must not exceed this and we note that with a 25 μm emulsion an alignment error of a 2° error can be tolerated. Thin emulsions give a greater statistical fluctuation in the number of developed grains per unit area and necessitate longer exposures.

Owing to the dense packing of grains in nuclear emulsions, it is difficult to obtain uniform development and in order to reduce the development rate so it is comparable with the diffusion rate of developer through the emulsion, development for high resolution work should be performed at about ice temperature. It is satisfactory to carry out developing in the main body of a domestic refrigerator running as cold as possible. Prior to development, the emulsion must be softened by soaking. The developer Kodak D19b keeps reasonably well when concentrated but goes off rapidly when diluted and the developer should therefore be changed daily. The stop bath and fixer can, however, be used repeatedly. Plates should be washed in cold, filtered tap-water for several hours, and then gently swabbed in running water with cotton wool to remove any remaining particles. Extreme care must be taken in keeping the solutions clean if good results are to be obtained. Plates should be dried under cover in air and not force-dried. The processing times of Ilford L4 emulsions are set out in Table 8.1.

Over a considerable density range, the Ilford emulsions are linear and it is therefore possible to compensate for incorrect exposure in the development procedure. While it is better to have an overexposed, underdeveloped plate than vice versa, underexposure can be compensated to a fair degree, though the range over which this is possible is limited. One should aim for a basic photographic density of unity from the perfect diffracting crystal in the thin crystal situation and then the dark 'direct' images give densities up to 2 or 3. In thick crystal anomalous transmission topographs a basic density of about 2 from the perfect crystal gives a good contrast from the (white) dynamical images but still enables increases in intensity to be observed. A general rule of thumb is to develop until a clear image is seen on the back of the plate under a red safelight.

Table 8.1 Processing times (in minutes) of Ilford L4 emulsions

	Thickness	
	50 μm	25 μm
Soak in filtered deionised water	10	5
Develop (1:3 D19b to deionised water)	15–60	12–30
Stop (1% glacial acetic acid in deionised water)	10	5
Fix (300 g sodium thiosulphate 30 g sodium bisulphite in 1 litre deionised water)	60	30
Wash (filtered tap-water)	upwards of 120	upwards of 120

On development, the grains of the L4 emulsion swell to about 0.25 μm. However, the granularity observed on the topograph when viewed under the microscope is not a result of the grain of the film but is statistical 'shot noise', arising from the statistical variation in the number of developed grains per unit area. This granularity can be reduced by increased exposure, but this strategy is limited by the low dynamic range of the emulsions. Lang[10] has considered the effect of this noise on resolution in some detail.

When using hard radiations (MoK_α and AgK_α) the resolution will be limited also by the length of the photoelectron tracks in the emulsion which are about 2 μm long with these radiations. Thus many factors conspire to limit topographic resolution to between 1 and 2 μm.

For high resolution topographs it is not usually satisfactory to magnify topographs with a photographic enlarger. Light scattering reduces the contrast of dislocations, and at the necessary magnification the depth of field is small. Experience shows that the best topographs come from the best (and most expensive) microscopes which have macro attachments for low resolution work. Magnifications over 500× are almost never used, while 30× to 60× provides a very useful range of enlargement. Photography should be performed on the microscope, and topographs are conventionally printed as positives.

8.6.2 *Real-time topography resolution*

The photographic method, while giving excellent resolution and contrast, is tedious and inconvenient in, for example, a semiconductor processing environment. It is also inapplicable to dynamic experiments in which the defect configuration varies. Real-time topography detectors fall into two categories – those employing direct conversion of X-rays to electrical signals and those using an X-ray to optical converter.

The former approach has been employed in the highest resolution direct viewing systems. As developed by Chikawa and colleagues[11] the diffracted X-ray beam passed through a beryllium window onto a thin PbO photocathode of a sensitive vidicon TV tube. The sensitivity of the device is finally limited by the target thickness, as a thin target absorbs only a small fraction of the X-rays, and in thick targets only a thin layer on the X-ray entrance surface side becomes photoconducting. Initially, Chikawa obtained a resolution of 30 μm, but subsequent developments in the electronics of such a camera have resulted in a resolution of 8 μm.[12] Unfortunately, the tube is sensitive to radiation damage and is not produced commercially in the high resolution form. Further, owing to the limited number of TV lines, the field of view of the camera at high resolution is necessarily small.

Indirect conversion devices have been reported with comparable spatial resolution[13] based on a thin fluorescent screen in the form of a microscope objective lens. This again has a small field of view and poor efficiency owing to the low stopping power of the phosphor screen. Commercial devices based on a channel plate intensifier coupled to a CCD camera usually give a spatial resolution of about 25 μm, which is adequate for many industrial applications where screening of substrates is required. However, only slow progress has been made in the past decade in resolving the conflict between the requirements of high spatial resolution, large field of view and high efficiency.

In order to see why these conflicts arise, we consider a flux I per unit area per unit time of X-ray photons incident on a detector of efficiency η. The number of photons N detected in an integration time τ in a square picture element of side ε is then

$$N = I\eta\varepsilon^2\tau \tag{8.1}$$

If two elements in the topograph have an intensity difference ΔI, then the difference in signal ΔN is

$$\Delta N = \Delta I\eta\varepsilon^2\tau \tag{8.2}$$

When written in terms of the contrast C, defined as $C = \Delta I/(2I + \Delta I)$, this becomes

$$\Delta N = \eta\varepsilon^2\tau 2IC/(1 - C) \tag{8.3}$$

Now the rms noise on the signal is $(2N + \Delta N)^{1/2}$ and the signal-to-noise ratio R is then given by

$$R = \Delta N(2N + \Delta N)^{-1/2} = (C\Delta N)^{1/2} \tag{8.4}$$

Thus

$$\varepsilon = \frac{R}{C}\left(\frac{1-C}{2\eta\tau I}\right)^{1/2} \tag{8.5}$$

This is a form of the Rose–de Vries equation and we see that it sets a fundamental limit on the spatial resolution of a quantum limited system. The spatial resolution is improved by increase of the incident intensity I, the detector efficiency η and the integration time τ. The former has been greatly improved by use of synchrotron radiation but there is a fundamental problem with improving the efficiency. To retain the high spatial resolution, phosphors or TV target regions must be thin and therefore stop only a small fraction of X-rays. Use of heavy, rare earth-based phosphors or phosphors deposited onto etched fibre-optic plates provide some ways of reconciling these fundamental contradictions. Note that for quality assurance applications, the use of integration times of the order of minutes may be perfectly acceptable and with PC-based image capture electronics and software, very high quality images can be reconstructed by digital integration.

8.6.3 *Optimal contrast and speed*

The intensity dI reaching the crystal due to a small area dX, dY of the X-ray source is

$$dI = P\,dX\,dY/HVD^2 \tag{8.6}$$

where P is the total X-ray tube power, H the dimension in the incidence plane, V the projected (vertical) dimension of the source normal to the incidence plane, and D the source to specimen distance. In the incidence plane only a fraction $\Delta\theta_B$ of the reflecting range is diffracted and the total diffracted intensity becomes

$$I \propto (P/HVD^2)\int_0^V dX \int_0^{D\Delta\theta_B} dY = P\Delta\theta_B/HD \tag{8.7}$$

Assuming a linear response of the film, the exposure time t for a stationary topograph is therefore

$$t = \left(k\Delta\theta_B P/HD\right)^{-1} \tag{8.8}$$

with k a constant of proportionality.

For a traverse topograph

$$T = \left\{k\Delta\theta_B \left(P/HD\right)\left(M/R\cos\theta_B\right)\right\}^{-1} \tag{8.9}$$

where R is the length of traverse and M the slit width.

In order to minimise the exposure time, we chose $M = H$, so the whole source is used. Then, neglecting the distance from the collimating slit to the specimen, we require for monochromaticity

$$H = M < D\Delta\theta_{\lambda_1\lambda_2} \tag{8.10}$$

where $\Delta\theta_{\lambda_1\lambda_2}$ is the angular separation between adjacent X-ray lines. Thus,

$$t = \left(C\Delta\theta_B\Delta\theta_{\lambda_1\lambda_2} \, P/H^2\right)^{-1} \tag{8.11a}$$

for the stationary topograph and

$$T = \left(C\Delta\theta_B\Delta\theta_{\lambda_1\lambda_2} \, P/HR\cos\theta_B\right)^{-1} \tag{8.11b}$$

for the traverse topograph.

These equations give the optimum conditions for a rapid high resolution topograph when only the resolution in the incidence plane is considered. However we must also consider the resolution normal to the incidence plane (Figure 8.4) and this is given geometrically by

$$\delta = VL/D \tag{8.12}$$

In order to separate direct and diffracted beams,

$$L > M/\tan 2\theta_B \tag{8.13}$$

That is

$$\delta_L = VL\Delta\theta_{\alpha_1\alpha_2}/H \tag{8.14}$$

Under optimum conditions the resolution normal to the incidence plane of a Lang topograph will be

$$\delta > V\Delta\theta_{x_1x_2}/\tan 2\theta_B \tag{8.15}$$

The importance of always placing the plate as close as possible to the specimen is clearly demonstrated. Substitution of typical values ($V = H$ and $\Delta\theta_{\alpha_1\alpha_2} = 2 \times 10^{-4}$) gives a resolution of $2\,\mu m$ when $L = 1\,cm$. As the value of $\Delta\theta_{\alpha_1\alpha_2}$ is a lower limit, corresponding to a Bragg angle of 6° we note that it is only with extreme care that the limits of resolution can be reached.

8.7   Contrast on X-ray topographs

8.7.1   *Crystals apparently perfect*

As a result of the relatively poor spatial resolution in X-ray topographs, there can be confusion as to whether uniform contrast results from a highly perfect or a highly imperfect crystal. There are, however, very important tests which can be applied, none more so than for the section topograph where the contrast from a perfect crystal is far from uniform.

Unlike the projection (or traverse) topograph, in section topographs of parallel-sided plates, interference, or thickness, fringes are observed (Figure 8.6(a)). They arise from dynamical interference between waves propagating within the Borrmann fan. As we saw above, the section topograph is formed when a ribbon beam passes into the crystal and the diffracted intensity in a stripe down the crystal is recorded on the photographic plate. As the divergence of the beam is large compared with the perfect crystal reflecting range we cannot make the assumption that the incident wave is a plane wave. This is unlike the situation in transmission electron microscopy, where the strong scattering results in a very large range of reflection and a plane wave approximation is usually valid. Provided that the width of the incident beam is small compared with the width of the Borrmann fan, to a first approximation the wave at the entrance surface can be considered as a spherical wave. As a plane wave outside the crystal excites two Bloch waves on opposite branches of the dispersion surface, the spherical wave excites the whole of the dispersion surface and wavefields propagate in all directions within the Borrmann fan (Figure 8.6(b)). As they are excited coherently, they preserve their relative phase and will therefore interfere. At a point on the exit surface R, two wavefields arrive, one from a point S on branch 2 and one from a point T on branch 1 of the dispersion surface (Figure 8.6(c)). As these wavefields lie on different branches of the dispersion surface, they have different wavevectors and there will be a phase difference between them at the exit surface. The phase difference clearly changes across the base of the Borrmann fan and across the section topograph of a parallel-sided, perfect crystal a series of interference fringes is seen (Figure 8.6(a)).

These fringes are known as 'Kato fringes'[14,15] and they provide important information on the dispersion surface shape. The density of the fringes is low at the centre of the pattern, resulting from the angular amplification of the angular deviation of the incident rays discussed by Authier.[16] The fringe density increases towards the margins of the section topograph image, as here the wavevector difference becomes greater for a smaller change in the normal to the dispersion surface. With increasing absorption, the fringe contrast becomes weaker, as only the branch 2 wavefield reaches the exit surface and in very thick crystals, the section pattern becomes narrow, with only the branch 2 wavefield in the centre of the dispersion surface reaching the exit and therefore no fringes are visible.

As the effect is a 'perfect crystal' effect, the presence of small defects, which may be too small or of too high a density to be imaged individually can be detected from the destruction of the phase relationships and the resulting absence of the Kato fringes. In crystals which have been irradiated or internally oxidised, small point defect clusters develop, the existence of which can be inferred from the loss of visibility in the section topograph thickness of Pendellösung fringes (Figures 8.6(d) and 8.6(e)). The depth resolution of section topography can be used to monitor the width of the denuded zone in silicon which has been intrinsically gettered to remove oxygen from the near-surface region used for device fabrication.[17,18]

As sketched in Figure 8.6(c), the surfaces of constant phase are on hyperbolic cylinders and thus in a parallel-sided plate the exit surface cuts these cylinders in straight lines. Hence parallel-sided fringes are observed. When the crystal is wedge-shaped, the exit surface cuts the hyperbolic cylinders along hyperbolae (Figure 8.7) and thus hyperbolic fringes are seen in the topograph.

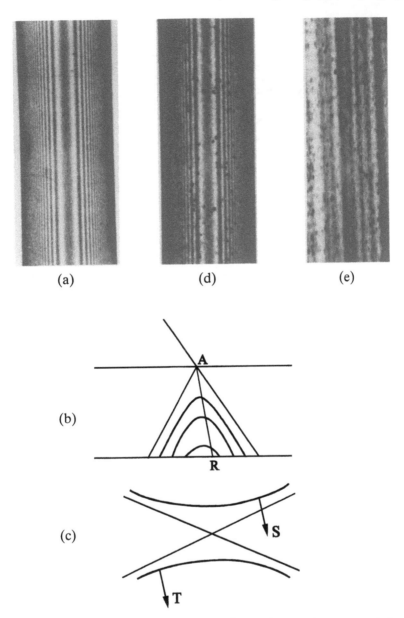

(a)   (d)   (e)

(b)

(c)

Figure 8.6 (a) Spherical wave thickness, or Kato, fringes in a section topograph from a perfect parallel-sided silicon crystal. (b) Schematic diagram of the section topograph geometry showing the surfaces of constant phase and the ray path AR of wavefields associated with tie-points S and T on the dispersion surface. (c) Dispersion surface construction showing tie-points S and T. (d) Experimental section topograph of the same crystal as in (a) showing loss of visibility of Kato fringes after annealing and precipitation of oxygen. (e) Complete loss of Kato fringes after further annealing. (Courtesy J. R. Patel)[19]

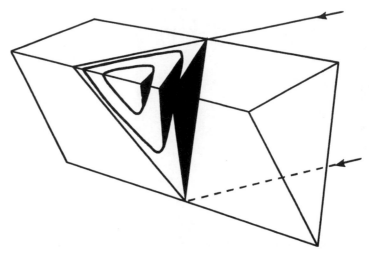

Figure 8.7 Schematic diagram of the surfaces of constant phase in a wedge-shaped crystal

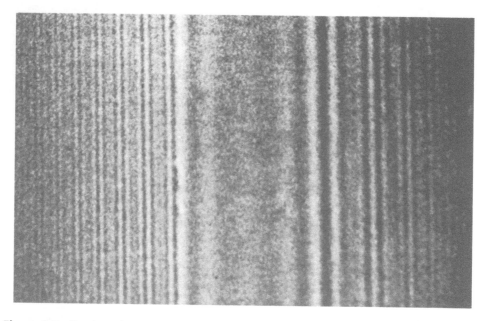

Figure 8.8 Beating of Kato fringes due to the presence of two states of polarisation in a section topograph of a crystal of diamond. CuK$_\alpha$ radiation. (Courtesy S. S. Jiang, University of Nanjing)

With reasonably high Bragg angle and a conventional generator, the Kato fringes show a periodic fading and spacing shifts due to the two states of polarisation having different wavevectors and thus different extinction distances (Figure 8.8). Section topograph thickness fringes have been used to make extremely precise measurements of structure factors.[20-22] Long-range strain is the most serious potential cause of error as this causes contraction of the fringes.

Section topographs provide valuable information on the direction of energy flow in the crystal. Although it may be shown that rays in the centre of the Borrmann fan have higher intensity than those in the margins, this does not give us the average intensity across the section topograph. The angular amplification A is defined as

$$A = d\theta/d(\Delta\theta) \tag{8.16}$$

where $d(\Delta\theta)$ is the angular divergence of an incident wave-packet and $d\theta$ is the corresponding angular divergence of the wavefields inside the crystal. We recall that the propagation direction of a wavefield is perpendicular to the dispersion surface at the associated tie-point. Defining the dispersion surface radius of curvature as R, it is then straightforward to show that

$$A = k\cos\theta_B/R\cos\theta \tag{8.17}$$

We see that very close to the Bragg condition, where the dispersion surface is highly curved, $R \ll k$ and the crystal acts as a powerful angular amplifier. A reaches 3.5×10^4 in the centre of the dispersion surface for silicon in the 220 reflection with MoK_α radiation. Far from the centre, the dispersion surface becomes asymptotic to the spheres about the reciprocal lattice points and A approaches unity. Thus when the whole of the dispersion surface is excited by a spherical wave, owing to the amplification close to the Bragg condition, the density of wavefields will be very low in the centre of the Borrmann fan and extremely high in the margins. Calculations of the intensity, averaged over the spherical wave thickness oscillations, compounding the intensity and density of wavefields, have been performed by Kato.[23]

The very important feature of the low absorption curve, visible experimentally in Figure 8.6(a), is that the diffracted beam intensity rises very sharply at the edges of the section pattern. It arises from the very high density of wavefields in these regions and its presence is a good indication that the crystal is highly perfect. Point defect clusters, for example from radiation damage, will destroy the phase relationships and destroy the margin effect. Under moderate absorption conditions ($\mu t \approx 2$), however, the perfect crystal profile is very similar to that of a crystal containing a large amount of damage and under such conditions topographs should interpreted with care.

8.7.2 Thickness fringes in traverse topographs

Takagi[24] has shown that the intensity on a projection topograph is given by the integrated intensity and that this is independent of the detailed shape of the wavefront. The phase relations are lost, and we can use the integrated plane wave intensity to calculate the intensity of the diffracted beam. For the perfect crystal with zero absorption the integrated diffracted intensity I_g^T is given by

$$I_g^T = (g\xi_g)^{-1}\int_{-\infty}^{\infty}\sin^2\left\{\pi\left(1+\eta^2\right)^{1/2}\Big/\xi_g\right\}\Big/\left(1+\eta^2\right)d\eta = \left(\pi/2g\xi_g\right)\int_0^{2\pi/\xi_g}J_0(\rho)d\rho \tag{8.18}$$

where J_0 is the zero-order Bessel function. This is plotted in Chapter 4 as Figure 4.23.

The term outside the integral shows that high-order, long extinction distance reflections give very low intensity while the integral determines how the intensity varies as a function of thickness. We note that the intensity oscillates with pseudo-period of ξ_g, the extinction distance. These oscillations are the thickness or Pendellösung fringes observed in a wedge crystal in Lang and white beam synchrotron radiation topographs. As the thickness increases, the amplitude of the oscillation decreases as $(\xi_g/t)^{1/2}$ and thus the visibility decreases. When absorption is included, the mean intensity decreases with thickness.

8.7.3 Crystals containing sub-grains and dislocations

As discussed earlier, there are two fundamental mechanisms for contrast in X-ray topographs. The first, orientation contrast, is appropriate to crystals in which the dislocations configure to form low-angle sub-grain boundaries. In Lang and Berg–Barrett topographs, orientation contrast occurs where part of the crystal is misoriented in such a way that diffraction cannot occur at the same time as that from the rest of the crystal. Zero intensity corresponds geometrically to the misoriented region (Figure 8.1) when the misorientation exceeds the divergence of the incident X-ray beam.

The second type of contrast is termed extinction contrast and results from the scattering power around the defect differing from that in the perfect crystal. There are essentially three types of extinction contrast, identified by Authier,[25] which can be seen schematically in Figure 8.9(a). The three images which are formed are termed the direct image (1), arising from the diffraction of X-rays which do not satisfy the Bragg condition in the perfect crystal, the dynamical image (2) which is formed from changes in intensity in the Bloch wavefields propagating through the perfect crystal, and the intermediary image (3), formed by interference between these wavefields and new wavefields created at the defect. These images are found in cases where the strain gradient is high, such as around a dislocation, and this leads to the region around the defect behaving as if it were the surface of the crystal. The

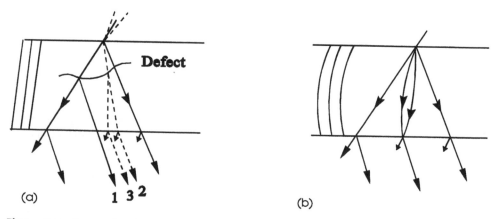

Figure 8.9 Types of extinction contrast. (a) Defect with high strain gradient. Three images, the direct (1), the dynamical (2) and the intermediary (3) are formed. (b) Low strain gradient where the wavefield rays paths are curved

wavefields decouple into their plane wave components and create new Bloch wavefields when the strain field becomes small again below the defect. A somewhat different type of dynamical image is formed when the strain gradients are small (Figure 8.9(b)). Then the wavefields do not decouple, but the tie-points migrate along the dispersion surface. As the propagation direction, that is the Poynting vector, is always normal to the dispersion surface, the rays paths are curved in real space. These two effects lead to redistribution of the energy between forward and diffracted beams, resulting in contrast on the topograph.

8.8 Theoretical tools for the interpretation of X-ray topograph contrast

8.8.1 *Small distortions*

When the deformation is small, the concept of a ray is retained and the crystal deformation is described by a continuously varying reciprocal lattice vector. This is equivalent to the situation in optics when light propagates through a medium of slowly varying refractive index. The original modes of propagation in the undeformed crystal retain their identity, in the X-ray case as Bloch wavefields, but in order to accommodate the deformation each mode undergoes a gradual transformation. Eikonal theories have been developed by Kato[26] and Penning and Polder.[27] Here, we follow the treatment of Tanner.[8]

In a distorted crystal, where the atomic displacement from the perfect crystal is given by the vector u, we can define a local reciprocal lattice vector g' by

$$g' = g - \mathrm{grad}(g.u) \tag{8.19}$$

Let us consider two points F and G between which the local reciprocal lattice varies from g' to $g'+dg'$. If the deformation is small the shape of the dispersion surface does not change and only a displacement of the hyperbolae results. We can consider this as a rotation about the origin of reciprocal space, and then the dispersion surface moves along the asymptote (Figure 8.10) by a vector given by v, where

$$\mathrm{v}.K_0 = 0 \quad \text{and} \quad \mathrm{v}.K_g = K_g.dg' \tag{8.20}$$

No movement of the dispersion surface takes place when $K_g.dg' = 0$ and thus for a local reciprocal lattice where the vectors dg' are always perpendicular to K_g, the wavefields propagate unchanged. Generally, the loci of points where dg' is perpendicular to K_g form surfaces along which the dispersion surface remains static and the associated wavefields unchanged. These surfaces divide the crystal up into domains which can be considered as perfect and are analogous to the surfaces of constant refractive index in the optical case. As each domain will be misoriented or dilated with respect to its neighbour, the wavevector K_0 changes by dK_0 on crossing the domain boundary. In analogy with the optical case this is given by

$$dK_0 = \gamma\,\mathrm{grad}(g'.K_g) = -\gamma\,\mathrm{grad}(K_g.\,\mathrm{grad}(g.u)) \tag{8.21}$$

As the fundamental equations of the dispersion surface are

$$K_0^2 = k^2(1+\chi_0) + k^2 C\chi_{\bar{g}} R \tag{8.22}$$

$$K_g^2 = k^2(1+\chi_0) + k^2 C\chi_g / R \tag{8.23}$$

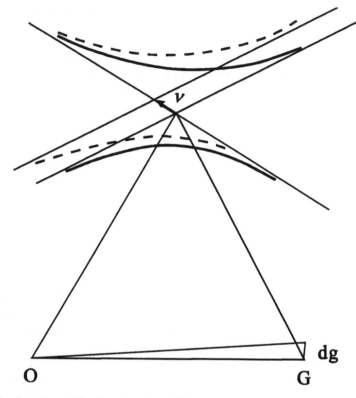

Figure 8.10 Rotation of the local reciprocal lattice vector in a distorted crystal, giving rise to a displacement of the dispersion surface hyperbola

where $R = |\mathbf{D}_g|/|\mathbf{D}_0|$ we derive, for $\chi_g = \chi_{\bar{g}}$

$$d\mathbf{K}_0 \cdot \left(\mathbf{K}_0 + R^2 \mathbf{K}_g \right) = -R^2 \mathbf{K}_g \cdot d\mathbf{g}' \qquad (8.24)$$

If $d\mathbf{r}$ is the path vector between F and G

$$d\mathbf{g}' = \left(d\mathbf{r} . \text{grad} \right) \mathbf{g}' \qquad (8.25)$$

Defining a vector $\mathbf{P}'$ parallel to the Poynting vector given by

$$\mathbf{P}' = \mathbf{K}_0 + R^2 \mathbf{K}_g \qquad (8.26)$$

we find

$$\gamma = -R^2 dl / \left\| \mathbf{P}' \right\| \qquad (8.27)$$

where dl is the path length along the ray trajectory given by

$$d\mathbf{r} = \mathbf{P}' \, dl / \left\| \mathbf{P}' \right\| \qquad (8.28)$$

Then

$$dR = \left(2R^2 dl / \left(k^2 C \chi_g \left| \mathbf{P}' \right| \right) \right) \left(\mathbf{K}_0 . \text{grad} \right) \left(\mathbf{K}_g . \text{grad} \right) \left(\mathbf{g} . \mathbf{u} \right) \qquad (8.29)$$

190

which may be written as

$$dR = \frac{2R^2 dl}{C\chi_g |\mathbf{P'}|} \frac{\partial^2 (\mathbf{g.u})}{\partial s_0 \partial s_g} \qquad (8.30)$$

Instead of considering the dispersion surface as a variable and the reciprocal lattice as invariant, it is usually easier to consider the reciprocal lattice as the variable. Then equation (8.30) determines the variation of the amplitude ratio of the reflected and transmitted components as the wavefield propagates through the crystal. The ratio R characterises a particular tie-point on the dispersion surface and if R varies the tie-point must migrate along the dispersion surface branch. This results in a change in the intensity of the transmitted and diffracted beams. Further, as the direction of energy flow, i.e. the ray direction, is everywhere normal to the dispersion surface, the ray direction varies as the tie-point migrates. The rays therefore propagate along curved paths.

The important parameter in determining the contrast is the integral along the ray path of the term $\{\partial^2/\partial s_0 \partial s_g\}(\mathbf{g.u})$. The ray path parameter β is defined as

$$\beta = (kC\chi_g)^{-1} \{\partial^2/\partial s_0 \partial s_g\}(\mathbf{g.u}) \qquad (8.31)$$

From equation (8.30) we see that the sign of the change dR is independent of the sign of R and depends only on β. The sign of R is opposite for the two branches of the dispersion surface but on both branches R increases with the deviation from the exact Bragg condition. Thus, for a given deformation, the tie-points on both branches of the dispersion surface migrate in the *same* direction. As has been experimentally verified by Hart and Milne,[28] the curvature of the ray corresponding to branch 2 is in the same sense as the curvature of the reflecting planes while the curvature of branch 1 rays is of opposite sense to the lattice curvature.

For the Eikonal theory to be valid, the distortions must be sufficiently small that the concept of a ray is retained. The criterion is that the radius of curvature of the reflecting planes does not exceed a critical value R_c, approximately equivalent to an angular rotation of the Bragg planes by half the reflecting curve width in an extinction distance. The critical radius of curvature is thus

$$R_c \approx g\xi_g^2 \qquad (8.32)$$

Although, Kato[29] has shown that the ray theory breaks down approximately $10\,\mu m$ from a dislocation core and we cannot apply it to compute the contrast of a dislocation in a quantitative way, it is possible to obtain a large amount of qualitative information.

In a thick crystal, as a result of the Borrmann effect, only the branch 2 wavefields close to the exact Bragg condition are present and the energy flow is parallel to the lattice planes. Let us consider first a defect such as a precipitate in the centre of a thick crystal and look at the contrast a long way from the defect (Figure 8.11(a)). On one side, the curvature is first positive and then negative while the converse is true on the opposite side of the defect. Thus the tie-point migration is in opposite sense on either side of the dislocation line but, provided the distortion is small so that the tie-points retain their identity, the migration above and below the defect exactly compensates on both sides of the defect. Close to the exact Bragg condition (which corresponds to our branch 2 anomalously transmitted wavefield) the angular amplification is extremely large (equation (8.17)) and hence very small distortions of the Bragg planes lead to appreciable changes in the direction of energy flow. The rays so deviated suffer additional absorption because the effective absorption coefficient is sensitive to the deviation from the exact Bragg condition. Thus there is a loss of intensity in the rays coming from both sides of the dislocation, and the long-range image of the defect appears white. Such contrast is commonly observed from dislo-

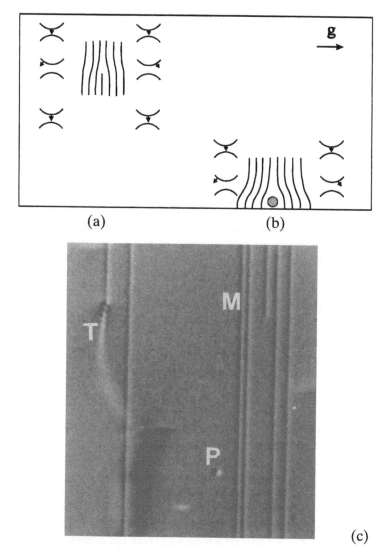

(a)   (b)

(c)

Figure 8.11 (a) Schematic diagram of tie-point migration in the long-range strain field of a dislocation in the centre, and a precipitate near the exit surface of a thick crystal. (b) Black–white contrast from misfit dislocations M and a precipitate P close to the exit surface of a thick crystal. Note that the contrast of the threading dislocation T is white, this being the dynamical image and oscillatory contrast; the intermediary image is seen only near the exit surface

cations in anomalous transmission topographs and a good example is the threading dislocation T in Figure 8.11(b).

When the defect is close to the surface of a thick crystal, however, we may observe opposite contrast on either side of the defect. Consider first the right-hand side of the precipitate in Figure 8.11(a). The ray path parameter β is first negative and the tie-point of the branch 2 wavefield migrates to the right, following the lattice plane curvature. At the surface the lattice planes emerge normally, and when the defect is close to the surface the lattice planes must rotate very rapidly. Thus the curvature is very large – too large for the Eikonal theory to be applicable – and the rays pass out of the crystal without further diffraction. Qualitatively, we find from equations (8.30) and (8.31) that the migration causes an increase in the intensity in the diffracted beam. Conversely, on the left-hand side of the defect we find decreased intensity. A precipitate close to the surface shows black–white contrast which is enhanced on the side of positive g when the surrounding lattice is under compression. An example of the black–white contrast due to the differing sense of lattice curvature around misfit dislocations and a precipitate close to the exit surface of a strongly absorbing crystal of GaAs is shown in Figure 8.11(b).

The black–white contrast reverses with the diffraction vector and also with the sense of the strain in the lattice. This is a useful means of determining the nature of a precipitate with a convenient 'rule of thumb'. On the side of positive g, if the contrast is enhanced, the lattice is under compression, if reduced, it is under tension.

8.8.2 *Large distortions*

When the distortion is high, we may obtain qualitative insights by retaining the concepts of the perfect crystal and treating the defective region as a crystal boundary. We then find that this leads to the concept of tie-points jumping from one branch to another when we match the wavefields across the interface. However, for quantitative calculations and simulation of dislocation images we need to go to the generalised form of the dynamical theory developed by Takagi.[21]

In the distorted crystal the wavevector $\mathbf{K}'_g$ differs from the ideal crystal wavevector by $\Delta\mathbf{g}$ where

$$\Delta\mathbf{g} = -\mathrm{grad}(\mathbf{g.u}) \tag{8.33}$$

and $\mathbf{u}$ is the atomic displacement. The β'_g parameter then is a function of position in the crystal. The electric displacement in the crystal has a form

$$\mathbf{D}(\mathbf{r}) = \sum_g \mathbf{D}'_g(\mathbf{r}) \exp(-2\pi\mathrm{i}(\mathbf{K}_g.\mathbf{r} - \mathbf{g.u})) \tag{8.34}$$

Takagi's equations are then

$$\partial D'_0/\partial s_0 = -\mathrm{i}\pi k C \chi_{\bar{g}} D'_g$$
$$\partial D'_g/\partial s_g = -\mathrm{i}\pi k C \chi_g D'_0 + 2\pi k \beta'_g D'_g \tag{8.35}$$

These may be combined to give two equations of the form

$$\partial^2 D'_0/\partial s_0 \partial s_g - \mathrm{i}2\pi k\beta'_g \partial D'_0/\partial s_0 + \pi^2 k^2 C^2 \chi_g \chi_{\bar{g}} D'_0 = 0$$
$$\partial^2 D'_g/\partial s_0 \partial s_g - \mathrm{i}2\pi k\beta'_g \partial D'_g/\partial s_0 + (\pi^2 k^2 C^2 \chi_g \chi_{\bar{g}} - \mathrm{i}2\pi k \partial \beta'_g/\partial s_0) D'_g = 0 \tag{8.36}$$

These have hyperbolic form, and in a perfect crystal can be solved analytically. In a distorted crystal they must be solved by numerical integration. Authier *et al.*[30] were the first to devise an algorithm for solution over a grid of points. If, in Figure 8.12, M is the point at which we require the wave amplitude and P and Q are neighbouring points at which the amplitudes are known,

$$D_0(M) = D_0(P) + p\left(-i\pi k C \chi_{\bar{g}}\right)D_g(P)$$
$$D_g(M) = D_g(Q) + q\left(-i\pi k C \chi_g\right)D_0(Q) + q\left(2\pi i k \beta_g\right)D_g(Q) \tag{8.37}$$

where $p = PM$ and $q = QM$ as defined in Figure 8.12. It is therefore straightforward, to derive the amplitudes at any point in the crystal by a step-by-step calculation.

Although the simulation is substantially slower than from the equivalent simulations of images in electron microscopy, section topographs can now be simulated in a matter of seconds on a fast personal computer. Just as rocking curve simulation can be used as a means of determining the microscopic strains from the scattered intensity, so simulation of topograph images can be used to determine the microscopic strain field around a defect and hence identify it. The match between simulation and experiment is extremely good, although in the section topograph the

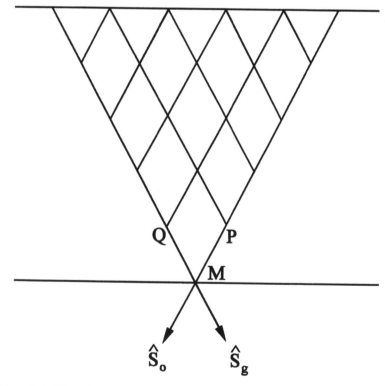

Figure 8.12 Grid filling the inverted Borrmann fan over which numerical integration of Takagi's equations can be accomplished

position of the defect does influence strongly the contrast (Figure 8.13). In all cases a very simple strain field of the form

$$u(r) = Cr/r^3 \qquad (8.38)$$

has been used to model the strain field around a precipitate. This radial strain field, based on isotropic elasticity theory is found to work very well for precipitates in silicon.

Epelboin[31] has given an excellent review of the procedures for calculating section topograph images and has more recently described a variable step algorithm which enables the grid size to be adjusted as the deformation increases in the vicinity of the defect.[32] A similar approach has been used in the simulation of the Bragg case double-crystal topographs by Spirkl *et al.*[33] More recently, Epelboin has given algorithms for parallel computation.[34] The diffraction process takes place solely in the incidence plane and therefore very rapid calculation can be achieved on parallel architecture machines by treating each slice independently and in parallel.

The wave theory equations essentially have the form of a multiple scattering theory based on plane waves. In the limit of small Bragg angles Takagi's equations reduce to the Howie–Whelan equations, well known to electron microscopists.

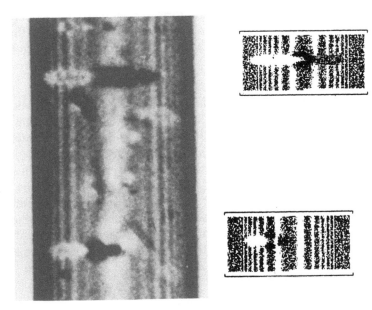

Figure 8.13 Experimental and simulated section topography images of hydrogen precipitates in silicon. All simulations were carried out using the same microscopic strain field, but the different position of the defects within the crystal results in substantially different image detail

8.9 Dislocation and stacking fault images in X-ray topographs

The contrast of defects in X-ray topographs has been recently reviewed by one of the authors[35] to which the reader is referred for further details and a bibliography. In this section we will confine the discussion to dislocation and stacking fault images, these being the most common examples of linear and two-dimensional defects.

8.9.1 *Images of dislocations*

Dislocation images in single-crystal X-ray topographs differ from those in electron micrographs primarily because the incident beam cannot be regarded as a plane wave. Of the three types of image characterised by Authier, the most common one seen in projection or white radiation synchrotron topographs is the direct image. This appears dark against the perfect crystal background on the topograph (Figure 8.14). The origin of this enhanced intensity lies in X-rays which are outside the range of diffraction of the perfect crystal. Although not diffracted by the perfect crystal, they will be in the deformed region around the dislocation. They thus suffer no primary extinction and the so-called direct image appears as an intense dark spot on the section topograph. In thick crystals the rays far from the Bragg angle contributing to the direct image suffer ordinary absorption and hence the direct image does not appear.

We can begin to understand the contrast mechanism by considering the crystal to be made up of three distinct regions, namely the perfect crystal above the defect, the perfect crystal below the defect, and the deformed region around the defect (Figure 8.15(a)). We set the limit of the deformed region as that where the effective misorientation $\delta(\Delta\theta)$ around a defect exceeds the perfect crystal reflecting range:

$$\delta(\Delta\theta) = -\left(k \sin 2\theta_B\right)^{-1} \partial(\mathbf{g.u})/\partial s_g \tag{8.39}$$

The region around the defect is assumed to diffract as a small mosaic crystal of thickness Δ, the X-rays diffracted from this region not being diffracted by the perfect crystal. Any dynamical diffraction effects associated with wavefield matching above and below the defect are neglected in this analysis. The integrated intensity for the Laue case (Figure 8.15(b)) varies linearly with thickness for small values of thickness, but as the thickness increases, so the gradient decreases and, in the absence of absorption, oscillates about a constant value. [When absorption is included, there is superimposed a gradual decay of intensity with thickness.] As the gradient of the I versus t curve is everywhere less than that at the origin, the intensity I_t due to material thickness t is always less than the sum of I_Δ and $I_{t-\Delta}$ from separate regions of thickness Δ and $t - \Delta$. In Figure 8.15(b), where $t/\xi_g \approx 3$, corresponding to the third thickness minimum, it is very obvious that

$$I_t < I_\Delta + I_{t-\Delta} \tag{8.40}$$

There is thus always enhanced intensity around the defect, the contrast being a maximum when the thickness correponds to the first thickness minimum, $t/\xi_g \approx 0.88$. For the Bragg case and zero absorption (Figure 8.15(c)) the intensity is independent of thickness beyond a crystal thickness of about an extinction distance ξ_g. Once

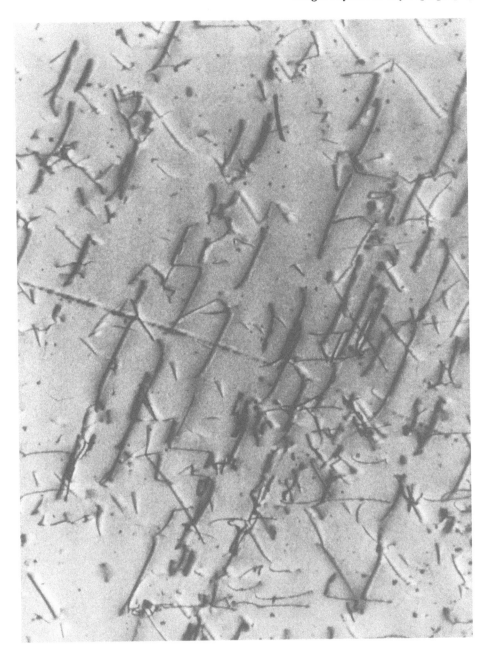

Figure 8.14 Images of dislocations in silicon under conditions of low absorption. AgK_α radiation. The dark images are the direct images. The white images are dynamical images

again we see that for crystal thickness greater than this value, equation (8.40) is valid and the defect shows enhanced intensity. In crystals less than an extinction distance in thickness, equation (8.40) becomes an equality, and defect images are not observed. This has important implications for studies of proteins where the extinction distances can be very long. The danger is that single-crystal topographs can appear perfect when in reality they may contain defects.

197

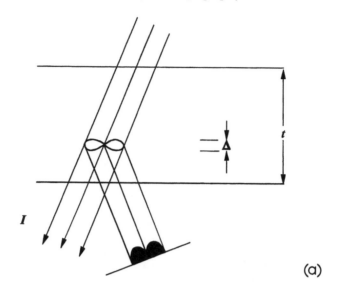

(a)

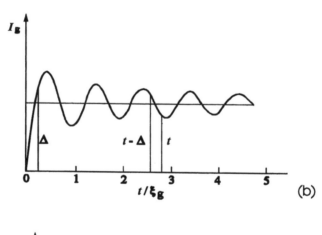

(b)

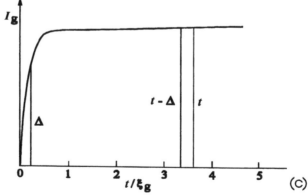

(c)

Figure 8.15 (a) Diagram showing the splitting of the crystal into perfect material above and below the defect and distorted material close to it. (b) Integrated intensity – Laue case (c) Integrated intensity – Bragg case

The width of the image can be deduced using this simple idea of contrast being formed when the misorientation around the defect exceeds the perfect crystal reflecting range $\delta\omega$. We consider the case of a screw dislocation running normal to the Bragg planes, where the line direction l coincides with the diffraction vector g. The effective misorientation $\delta\theta$ at distance r from the core is

$$\delta\theta = b/2\pi r \tag{8.41}$$

The width of the dislocation image D, which is twice the value of r for which $\delta\theta = \delta\omega$ and is thus

$$D = b/(\pi\delta\omega) \tag{8.42}$$

As, for a symmetric reflection, $\delta\omega = 2/g\xi_g$, we have $D = gb\xi_g/2\pi$. This result can be generalised to:

for screw dislocations $\quad D \approx g.b\,\xi_g/2\pi$ (8.43a)

for edge dislocations $\quad D \approx g.b\,\xi_g/\pi$ (8.43b)

Direct images of edge dislocations are approximately twice as wide as those for screw dislocations of the same $g.b$ value and this provides a useful method of checking Burgers vector assignment. As the extinction distance is on the scale of micrometres, dislocation images widths are of that order. The fact that this is several orders of magnitude greater than those of transmission electron micrographs is a fundamental limitation, set by the weak scattering, and hence high strain sensitivity, of X-rays in crystals. In terms of the structure factor F_g we have

$$D \approx constant.\left[\lambda F_g\right]^{-1} \tag{8.44}$$

and the dislocation image width goes down for increasing wavelength λ and strength of the reflection.

Equations (8.43) suggest that the image width is zero when $g.b = 0$. This is of course the classic criterion, originally applied to transmission electron microscopy, where the effective misorientation of the distortion around the dislocation is zero. As the Bragg planes are not distorted or tilted the dislocation is invisible in that reflection. Strictly, this criterion is that both $g.b = 0$ and $g.bxl = 0$ and, except when the dislocation runs parallel to a high symmetry axis, it is valid only for isotropic elasticity. Nevertheless, the contrast is often weak for just $g.b = 0$ and this enables b to be determined on purely geometric grounds by finding two reflections in which the dislocation is invisible or very weak.

Miltat and Bowen[36] showed that direct images can be synthesised from the cylinders of misorientation drawn around a dislocation line using continuum elasticity theory. The image full width can be calculated from the projected width circumscribed by the contour where $\delta(\Delta\theta)$ is equal to α times the reflecting range, with $\alpha \approx 1$. The intensity may be calculated from the volume of mosaic crystal contained within the cylinder. Unfortunately, the value of α is a function of the reflection used and the procedure cannot be used quantitatively to simulate the direct image. It does, however, provide a very rapid method for defect identification in association with other information. (Satisfactory simulation of the direct image has not yet been achieved, even in the case of the section topograph.) The model explains the significance of the bimodal profile of dislocations when $g.b > 2$ as here lobes of opposite misorientation on either side of the defect give rise to two (diverging or converging) components of the image. Such double images are readily seen at synchrotron radiation sources where the divergence effect is enhanced by the large specimen-to-plate distance.

An important feature of the direct image in the section topograph (Figure 8.16) is that the position of the direct image corresponds to the depth of the defect within the crystal. In the section topograph it is localised (Figure 8.16), unlike the dynamic and intermediary images which in the example here are spread out in a 'banana' shape. The localisation of the direct image has practical application in the control of intrinsic gettering of impurities in silicon. When this process is undertaken, a high density of precipitates is formed in the centre of the wafer, but a highly perfect region arises at the surface. As modern devices are shallow, this is a highly satisfactory production procedure. Under conditions of low absorption, section topography can be used non-destructively to monitor the width of these denuded zones at the surface, as the position of the direct image across the section topograph corresponds directly to the depth of the precipitate in the crystal.

Dynamical images, although seen in thin crystals when the dislocation lies close to the entrance surface and normal to the incidence plane (Figure 8.14), are generally important in thick or high atomic number crystals where the product μt is significantly greater than unity (Figure 8.17). They are less intense than the background and appear white on the plates. In thick crystals where only the branch 2 wave at the centre of the Borrmann fan remains, the presence of the dislocation causes scattering of intensity out of that wavefield into newly created wavefields. These suffer higher absorption and thus the dislocation casts a shadow in both diffracted and transmitted beams. In general the dynamical image is diffuse except where the dislocation is close to the exit surface of the crystal. This broadening of the image with distance from the exit surface can be clearly seen in Figure 8.11(b). Under extremely high absorption conditions, the dynamical images sharpen, as only those wavefields propagating parallel to the Bragg planes reach the exit surface. The intensity loss is then a sharp projection of the deformed region around the defect.

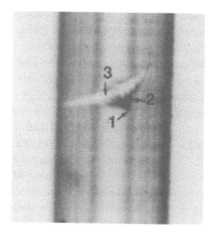

Figure 8.16 Section topograph of a dislocation in silicon. The dislocation cuts the Borrmann fan obliquely, resulting in displacement of the direct, intermediary and dynamical images. We note that the dynamical image, unlike the direct image, is not localised. 1, direct image; 2, intermediary image; 3, dynamical image[37]

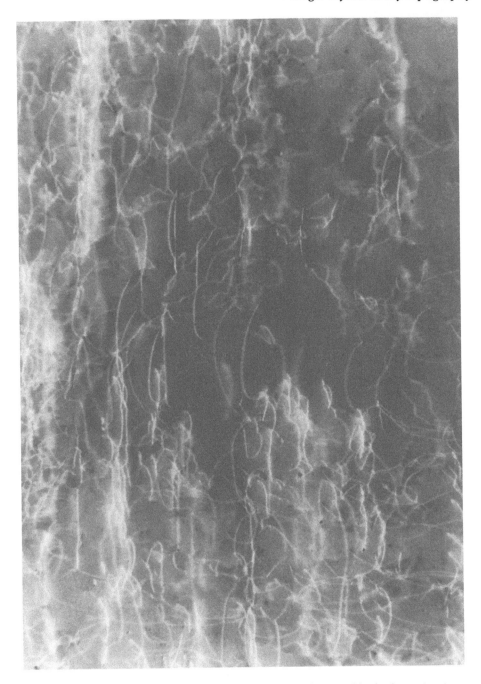

Figure 8.17 Dynamical images of dislocations under conditions of high absorption in GaAs. Note that the images appear light due to the loss of intensity around the defect

In some cases interference can be observed between the newly created wavefields leading to oscillatory contrast in the region of the dislocation close to the exit surface (the intermediary image). Only when the extinction distance is short are such effects usually observed, as anomalous absorption quickly damps out the

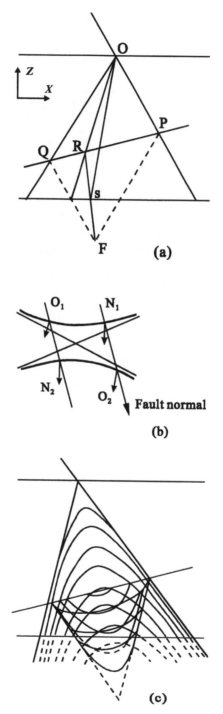

Figure 8.18 (a) Ray paths of original and new wavefields in a crystal containing an inclined stacking fault. (b) Tie-points O_1 and O_2 corresponding to original wavefields. Tie-points N_1 and N_2 corresponding to new wavefields. (c) Surfaces of constant phase. (d) Section topographs of an inclined stacking fault in diamond, (left) AgK_α radiation, (right) CuK_α radiation showing the periodic fading due to the two polarisation states. 111 reflection. Image height 0.85 mm. (Courtesy S. S. Jiang, University of Nanjing)

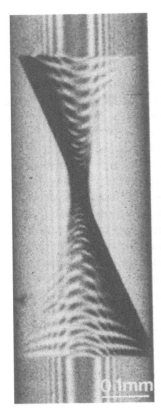

(d)

Figure 8.18 (*cont.*)

branch 2 waves within a short distance from the exit surface. Where they intersect the exit surface, dislocations often exhibit a black–white contrast from the surface relaxation which can be interpreted using the Penning–Polder theory along the lines discussed in section 8.8.1. Owing to divergence of the wavefields in the crystals, dynamical images have best contrast when the dislocation is close to the exit surface.

Intermediary images are usually blurred out by the traversing but often significantly affect the shape of the direct image leading to asymmetry of the profile. In suitable circumstances the intermediary image is well separated and, as in the section topograph, shows oscillatory contrast. The splendid intermediary images of Authier[23] are rare, and more often the overlap of direct and intermediary images leads to oscillatory contrast of an inclined dislocation.[32]

8.9.2 *Contrast of stacking faults in section topographs*

Stacking fault contrast in section topographs can be most easily understood by expanding the incident spherical wave as a Fourier expansion of plane waves using the method of Kato. It is also possible to use Takagi's theory to calculate the contrast, but the physical insight is lost.

203

We consider in Figure 8.18(a), a stacking fault inclined to the crystal faces. In the plane of incidence, the fault cuts the extrema of the Borrmann fan at P and Q. Above the fault two wavefields corresponding to the tie-points O_1 and O_2 in Figure 8.18(b) propagate in the direction OR. On crossing the fault they decouple into their plane wave components, but on re-entering the perfect crystal below the fault they immediately excite new wavefields. The boundary conditions can be applied to the plane wave components in a similar manner to those at a vacuum interface provided that the phase difference introduced by the fault is included. We see from Figure 8.18(b) that four waves now exist in the crystal. Two of these correspond to the original tie-points O_1 and O_2 and two to newly created Bloch wavefields corresponding to the newly excited tie-points N_1 and N_2. From the symmetry implied by the continuity of the normal component of wavevector, we see that the direction of propagation of the new wavefields N_1 and N_2 is the same. At a point S on the exit surface the intensity is made up of contributions from the newly created wavefields travelling in the direction OS. Explicitly, the intensity comprises three components, I_1, I_2 and I_3 where I_1 represents the interference between the original wavefields which gives rise to the section topograph thickness (or Kato) fringes, I_2 corresponds to the interference between the new wavefields and I_3 gives the contribution due to interference between original and new wavefields. In the case of a highly absorbing crystal, the expression for I_3 dominates as the contrast arises from the interference of waves associated with tie-points O_1 and N_1. These are both on branch 1 and are thus only weakly absorbed.

In the region OPR interference of the original wavefields occurs and the fronts of equal phase, shown in Figure 8.18(c), are hyperbolic. In the triangle QPF the newly created wavefields propagate and the phase fronts are also hyperbolae, equivalent to the phase fronts in OPQ reflected about PQ. In addition, the flat, *non-hyperbolic* phase fronts due to the interference of the two systems occur in QPF. These are shown as the additional contour in Figure 8.18(c). The newly created wavefields converge to a focus F and we note that F = OP and PF = OQ. As the line of intersection PQ moves through the crystal, so the position of the focus F varies. In the particular case where PQ is parallel to the crystal surface, it is easy to see that when PQ lies exactly half-way through the crystal, F is at the exit surface. Quite generally there is a position where no fringes from I_2 and I_3 are observed and the resulting section topograph from an inclined fault has the delightful 'hour-glass' shape illustrated in Figure 8.18(d). The flat fringes due to interference between original and new wavefields have twice the periodicity of the hyperbolic fringes.

In non-absorbing crystals, the hour-glass contains two sets of fringes, one of each periodicity, but in thicker crystals the I_2 term becomes vanishingly small. The contrast of the I_3 fringes is always higher than that of the I_2 fringes, and it is these non-hyperbolic fringes which are usually observed experimentally. A total of t/ξ_g flat fringes is found. Fringe fading arises from beating between the two states of polarisation, an effect not seen with synchrotron radiation.[38]

Under the classification of Amelinckx, stacking faults give rise to α fringes. The structure in the region below the fault is derived from that above by a simple translation **r** parallel to the fault. In contrast, across a twin boundary the region below the fault is related by a similar displacement vector **r** which increases linearly from the boundary and the fringes thus produced are known as δ fringes.

The term I_3 is sensitive to the sense of the phase difference, and examination of the contrast of the first fringe from the exit surface enables the sense of α and hence the nature of the fault, whether extrinsic or intrinsic, to be determined. The magnitude of **r** can be determined from the reflections in which the fault is invisible. This occurs whenever **g.r** = 0 or n where n is integral.[39,40]

Fringes are seen in traverse topograph images of stacking faults, but in general the contrast is weaker. Other planar defects such as magnetic domain boundaries, ferroelectric and ferroelastic domain boundaries, growth sector boundaries and growth bands themselves all give rise to images in X-ray topographs. The contrast may arise either from the elastic strains at the interface or the phase difference introduced into the wavefields on crossing the boundary. In many cases the contrast is complex and may be difficult to simulate.

The contrast of growth bands, which arise from changes in lattice parameter due to fluctuations in impurity levels, is particularly valuable in determining the growth history of crystals. No contrast is observed when **g.n** = 0, **n** being a vector normal to the growth front.

8.10 Summary

X-ray topography is the X-ray analogue of transmission electron microscopy and as such provides a map of the strain distribution in a crystal. The theory of image formation is well established and image simulation is thus a powerful means of defect identification. Despite a reputation for being a slow and exacting technique, with modern detector technology and care to match spatial resolution of detector and experiment, it can be a powerful and economical quality-control tool for the semiconductor industry.

References

1. L. G. SCHULZ, Trans. A.I.M.E., **200**, 1082 (1954).
2. A. GUINIER & J. TENNEVIN, Acta Cryst., **2**, 133 (1949).
3. A. R. LANG, J. Appl. Phys., **29**, 597 (1958).
4. U. BONSE & E. KAPPLER, Z. Naturforschung, **13a**, 348 (1958).
5. C. S. BARRETT, Trans. A.I.M.E., **161**, 15 (1945).
6. W. F. BERG, Naturwissenschaften, **19**, 391 (1931).
7. A. R. LANG, Acta Cryst., **12**, 249 (1959).
8. B. K. TANNER, X-ray diffraction topography (Pergamon, Oxford, 1976).
9. D. K. BOWEN & C. R. HALL, Microscopy of materials (Macmillan, London, 1975).
10. A. R. LANG, in: Modern diffraction and imaging techniques in materials science, eds. S. AMELINCKX, R. GEVERS & J. VAN LANDUYT, Vol. 2 (North-Holland, Amsterdam, 1970) p. 623.
11. J.-I. CHIKAWA, I. FUJIMOTO & T. ABE, Appl. Phys. Lett., **21**, 295 (1972).
12. S. SUZUKI, M. ANDO, K. HAYAKAWA, O. NITTONO, H. HAHIZUME, S. KISHINO & K. KOHRA, Nucl. Inst. Meth., **227**, 584 (1984).
13. W. HARTMANN, in: X-ray optics, Topics in Applied Physics, Vol. 22, ed. H.-J. QUEISSER (Springer-Verlag, Berlin, 1977) p. 191.
14. N. KATO, in: Crystallography and crystal perfection, ed. Ramachandran (1963), p. 153.
15. N. KATO, in: Introduction to X-ray diffraction, eds. L. G. AZAROFF, R. KAPLOW, N. KATO, R. J. WEISS, A. J. C. WILSON & R. A. YOUNG (McGraw Hill, New York, 1974) p. 425.

16. A. AUTHIER, Phys. Stat. Sol., **27**, 77 (1968).
17. T. TUOMI, M. TILLI & O. ANTTILA, J. Appl. Phys., **57**, 1384 (1985).
18. A. J. HOLLAND & B. K.TANNER, J. Phys. D: Appl. Phys., **28**, A27 (1995).
19. J. R. PATEL, J. Appl. Phys., **44**, 3903 (1973).
20. M. HART & A. D. MILNE, Acta Cryst., **A25**, 134 (1969).
21. N. KATO, Acta Cryst., **A25**, 119 (1969).
22. P. ALDRED & M. HART, Proc. R. Soc. Lond., **A322**, 233 and 239 (1973).
23. N. KATO, Acta Cryst., **13**, 349 (1960).
24. S. TAKAGI, J. Phys. Soc. Japan, **26**, 1239 (1969).
25. A. AUTHIER, Adv. X-ray Analysis, **10**, 9 (1967).
26. N. KATO, J. Phys. Soc. Japan, **18**, 1785 (1963).
27. P. PENNING & D. POLDER, Philips Res. Reports, **16**, 419 (1961).
28. M. HART & A. D. MILNE, Acta Cryst., **A31**, 425 (1971).
29. N. KATO, J. Phys. Soc. Japan, **18**, 1785 (1963).
30. A. AUTHIER, C. MALGRANGE & M. TOURNARIE, Acta Cryst., **A24**, 126 (1968).
31. Y. EPELBOIN, Mater. Sci. Eng., **73**, 1 (1985).
32. C. A. M. CARVALHO & Y. EPELBOIN, Acta Cryst., **A49**, 467 (1993).
33. W. SPIRKL, B. K. TANNER, C. R. WHITEHOUSE, S. J. BARNETT, A. G. CULLIS, A. D. JOHNSON, A. KEIR, B. USHER, G. F. CLARK, W. HAGSTON, C. R. HOGG & B. LUNN, Phil. Mag. A, **70**, 531 (1994).
34. Y. EPELBOIN, J. Appl. Cryst., **29**, 331 (1996).
35. B. K. TANNER, in: X-ray and neutron dynamical diffraction: theory and applications, eds. A. AUTHIER, S. LAGOMARSINO & B. K. TANNER (Plenum Press, New York, 1996).
36. J. MILTAT & D. K. BOWEN, J. Appl. Cryst., **8**, 657 (1975).
37. G. S. GREEN, N. LOXLEY & B. K. TANNER, J. Appl. Cryst., **24**, 304 (1991).
38. G. KOWALSKI, A. R. LANG, A. P. W. MAKEPEACE & M. MOORE, J. Appl. Cryst., **22**, 410 (1989).
39. A. AUTHIER, Phys. Stat. Sol., **27**, 77 (1968).
40. A. AUTHIER & J. R. PATEL, Phys. Stat. Sol (a), **27**, 213 (1975).

9

Double-crystal X-ray Topography

In this chapter we discuss double-crystal topography, in which we obtain a map of the diffracting power of a crystal compared to that of a reference. We first treat the principles and geometries, the mechanisms of image contrast and resolution and the use of laboratory and synchrotron radiation. We then discuss applications: wafer inspection, strain contour mapping, topography of curved crystals.

9.1 Introduction

As we saw in the last chapter, an X-ray topograph is a two-dimensional map of the scattered intensity distribution across a diffracted beam. Thus simply by placing a piece of photographic film or a direct viewing TV detector in the diffracted beam from the specimen in a high resolution diffraction experiment, one may take a simple X-ray topograph directly. For many purposes this simple procedure is very informative. For example, the presence of strong bands of intensity running at right-angles in a topograph taken from a (001)-oriented epitaxial layer immediately shows the presence of relaxation (Figure 9.1). In many cases this proves to be much easier than analysing symmetric and asymmetric reflections to determine the relaxation quantitatively. At this level of resolution, fast film, image plates or a direct read-out TV detector is satisfactory. However, if high spatial resolution or high strain sensitivity is required, rather more care must be taken in the design of the experiment in order for these parameters to be optimised. In this chapter we discuss the requirements for taking high quality double-crystal topographs and also the interpretation of the images obtained.

9.2 Principles and experimental geometries

Double-crystal topography was first performed in the late 1950s[1,2] but never then reached the level of usage that the single-crystal techniques achieved. The simplest experimental scheme to understand is that of the $(+n,-n)$ non-dispersive setting shown in Figure 9.2(a). Although the wavelength diffracted from different points across the crystal may vary, the beam emerging from the reference crystal is of very small angular divergence. We have seen in Chapter 2 that a very small rotation of the specimen is required to go right through the rocking curve. Thus any local

Figure 9.1 Double-crystal X-ray topograph of a heavily relaxed (001)-oriented GaAsSb layer on GaAs showing the characteristic 'tweed' contrast from two sets of misfit dislocations running parallel to the $\langle 110 \rangle$ directions

distortion of the specimen around a defect will result in loss of intensity and an image of the defect will be formed on the photographic plate. As the rocking curve widths have been shown to be very narrow, this implies a very high strain sensitivity of the $(+n,-n)$ non-dispersive setting. Two important things should be noted about this setting. If a point source is used with a conventional generator, two bands of intensity are observed across the topograph, corresponding to the $K_{\alpha 1}$ and $K_{\alpha 2}$ wavelengths. This may be adequate, but does not give a large area topograph with uniform illumination. If a line source is used, with the line focus being in the dispersion plane, then a uniform intensity will be recorded across a strip correspond-

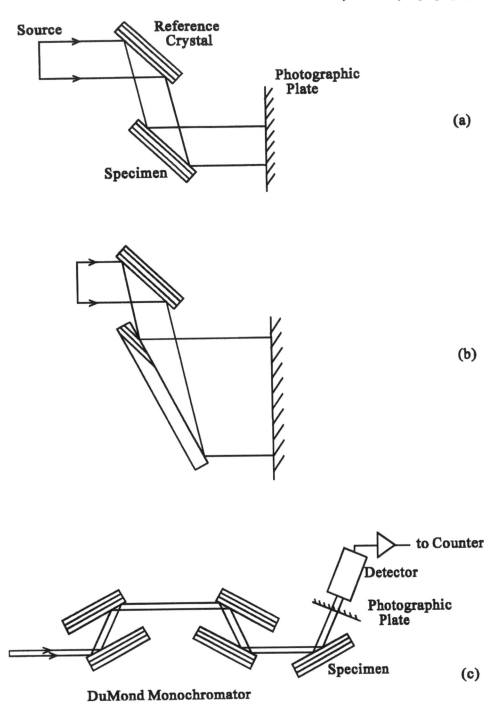

Figure 9.2 (a) Simple (+n,–n) double-crystal topography arrangement in the parallel setting. (b) Use of asymmetric reflection to expand illuminated area and eliminate image doubling. (c) Double-crystal topography with a standard duMond-Hart monochromator system

ing to the width of the source. However, because the $K_{\alpha 1}$ and $K_{\alpha 2}$ rays make different directions in space, double images will be observed of any defect. This may not be a problem in principle, but it does confuse interpretation.

To avoid this problem, an asymmetric reflection is often used from the specimen as shown in Figure 9.2(b). This permits the $K_{\alpha 2}$ beam to be removed by a slit between the specimen and the reference crystal, but still enables a large area of specimen to be imaged. This arrangement is also appropriate when a duMond-style beam conditioning monochromator is used. Convenient reflections with a high asymmetry factor for CuK_α radiation are the 224 and 044 for GaAs and InP and 113 for Si. The advantage of this arrangement is that high resolution topography can be performed without modification of the standard high resolution diffraction experimental arrangement (Figure 9.2(c)). Note that by using an asymmetric reflection to expand the beam from the reference crystal as well as the specimen, an expansion of up to several hundred times may be obtained. This provides a means of taking double-crystal topographs of very large area wafers. Hung Liu at Texas Instruments in Dallas has recorded full topographs of 5″ Si wafers using this method.

An alternative method for removing the double images is to use the dispersive setting, where the first and second crystals are not of the same material. For example, using the 004 reflection from a Si reference crystal with the 004 InP from the specimen will provide enough dispersion to diffract from only the $K_{\alpha 1}$ line. This inclusion of dispersion also reduces the extent of the defect image, as it reduces the sensitivity to strain.

9.3 Defect contrast in double-crystal topographs

Most of the features observed in double-crystal topographs can be interpreted by means of a simple geometrical theory first given by Bonse[3] in 1962. Let us assume that the shape of the rocking curve is triangular and that the specimen is oriented such that it is set on the flank of the rocking curve. Then it is very easy to see that the fractional change in intensity $\delta I/I$ is given by

$$\delta I/I = K\left\{\left(\tan\theta_B\right)\delta d/d\right\} \pm \mathbf{n}_t.\mathbf{n}_g\,\delta\theta \tag{9.1}$$

where K is the slope of the rocking curve, $\delta d/d$ is the fractional change in Bragg plane spacing, $\delta\theta$ is the tilt around the defect and $\mathbf{n}_t$ and $\mathbf{n}_g$ are unit vectors normal to the incidence plane and parallel to the defect misorientation axis respectively. The second term on the right-hand side is in fact the component of tilt in the dispersion plane. The sense of the contrast change depends on the angle that the specimen makes with the reference crystal. When the specimen is set so the rocking curve slope is positive (position (A) in Figure 9.3) a positive dilation results in increased intensity. When the slope K is negative (position (B) in Figure 9.3), then there is a reduced intensity for a positive dilation. When the specimen is set at the top of the rocking curve, the intensity is always reduced irrespective of the sense of dilation or tilt. Figure 9.4 shows examples of topographs of dislocations in InP set at points A and B respectively on the rocking curve. The complementary contrast of the defects in the two topographs is striking. The long sharp images of scratches, and that of the scratch A do not exactly reverse contrast, indicating that the strains here

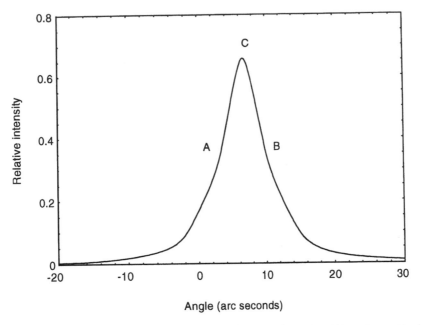

Figure 9.3 Schematic diagram of the positions of positive/negative (A) and negative/positive (B) contrast on either flank of the rocking curve. At the rocking curve maximum (C), the defect always results in an intensity loss, i.e. negative contrast

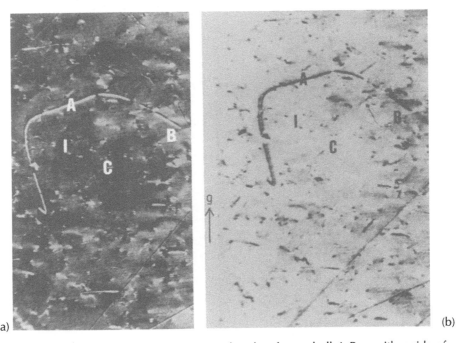

(a)   (b)

Figure 9.4 Experimental Bragg case topographs taken from a bulk InP on either side of the rocking curve: (a) low angle, (b) high angle[4]

are so large as to lead to departure from the roughly linear region on the rocking curve flank.

A full simulation of the defect image may be achieved by solving numerically Takagi's equations. This was first done for the double-crystal arrangement in the Bragg case by Riglet *et al.*[5] and more recently by Spirkl *et al.*[6] In most cases, the full dynamical theory simulation predicts a contrast reversal in the same way as the simple geometrical theory. However, the full dynamical theory does show that there are situations where, as with the experiment, the contrast does not simply reverse on going across the rocking curve. Figure 9.5 shows one such example. This simulated 'rocking topograph' is a simulation of the image of a misfit dislocation parallel to the surface of the crystal, with the offset from the Bragg position varying along the dislocation line length. As we pass from the bottom to the top of the image, we pass

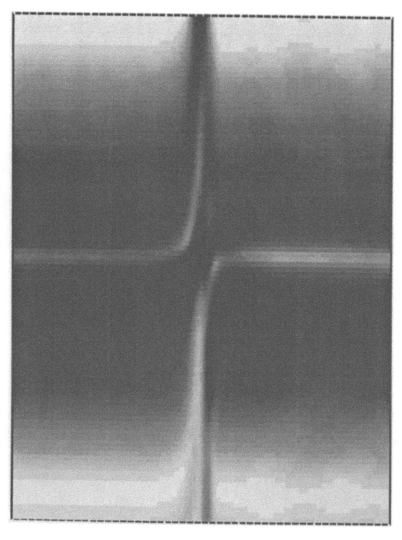

Figure 9.5 Simulated 'rocking topograph' of misfit dislocations parallel to the surface of a (001)-oriented GaAs wafer. Dislocation line runs up the page

through the rocking curve. Physically this corresponds to a twisted sample.[7] We note the reversal of contrast from the top to the bottom of the image. The case simulated is for misfit dislocations at the interface between a GaAs substrate and an InGaAs epilayer, but with a thick GaAs capping layer. This capping layer results in interference contrast which results in complex contrast at the dislocation. While the geometrical theory provides a valuable first step, for detailed interpretation of the microscopic strain field, full dynamical theory simulations are necessary.

9.4   Spatial resolution

The geometrical resolution normal to this dispersion plane in a double-crystal topograph is given in exactly the same way as for a single-crystal topograph. As for the single-crystal topograph, the spatial resolution δ is approximately given by

$$\delta = \frac{hb}{a} \qquad (9.2)$$

where h is the source size, b is the specimen to film distance and a is the source to specimen distance. For high spatial resolution, therefore, the film or detector must be placed as close as possible to the specimen. In the laboratory environment this effectively limits the spatial resolution to a few micrometres. However, this is normally perfectly adequate because the high strain sensitivity of the double-crystal techniques leads to very wide images from isolated defects. By reference to equation (9.1), it can be seen that when the rocking curve width becomes very narrow, the rocking curve gradient becomes very high. A large contrast change is then observed for a small misorientation or dilation. As the strain field around a dislocation falls off inversely with distance, the higher the strain sensitivity (larger value of K) the further is the distance away from the defect at which significant contrast is formed. In highly perfect crystals of Si containing a very small number of dislocations, the long-range nature of the strain field has been revealed, showing that the strain field extends throughout the whole specimen.

As seen in the last chapter, the image width is easy to quantify for a screw dislocation, where the diffraction vector is parallel to the dislocation line. Around a screw dislocation, the misorientation $\delta\theta$ at a distance r from the core is given by

$$\delta\theta = \frac{b}{2\pi r} \qquad (9.3)$$

The width of the image $D = 2r_0$ where r_0 is the value of r corresponding to a misorientation equivalent to the full width at half height maximum of the rocking curve $\delta\omega$. Thus, again,

$$D = \left\{b/2\pi\right\}/\delta\omega \qquad (9.4)$$

We note that if the rocking curve is approximated to a triangle, i.e. it has a constant slope, equation (9.1) can be used to calculate the intensity of the dislocation image in the topograph. As for the $(+n,-n)$ case,

$$\delta\omega = const.\lambda^2 F_h / \sin 2\theta_B \qquad (9.5)$$

for small Bragg angles,

$$D \approx const.\left[\lambda F_h\right]^{-1} \qquad\qquad (9.6)$$

Thus as for the single-crystal topograph, for a given reflection, the dislocation image width goes up with falling wavelength and for a given wavelength, the dislocation image width goes up with falling structure factor. Images using quasi-forbidden or high-order reflections are therefore intrinsically wide. This discussion shows that the higher the intrinsic sensitivity to strain, the wider are the images of defects.

The above discussion assumed that the reference and specimen crystals were of the same material and in the same diffraction conditions, i.e. the $(+n,-n)$ geometry. If the reference crystal is different from the specimen, then the distance from the specimen at which the image forms changes. If the reflecting range of the exit beam from the reference crystal is very large compared with the incident reflecting range of the specimen, then the image forms at a distance determined by the incident reflecting range width of the specimen. If, on the other hand, the reflecting range of the exit beam of the reference crystal is small compared with the incident reflecting range of the specimen, it is the reflecting range of the reference crystal which determines the image width. Thus the images of misfit dislocations observed in the 224 reflection from a (001)-oriented InGaAs layer on GaAs are wider and sharper if the 333 reflection is used for the reflection from a silicon reference crystal rather than the 111 reflection. In the former case, the lower divergence results in the incident beam not satisfying the Bragg condition over a greater volume of crystal around the defect and giving a better signal to noise in the image.

Use of grazing incidence has a beneficial effect on the image contrast, as well as having the effect of expanding the beam area. If the reference crystal has a grazing incidence beam, the exit beam divergence is reduced by $\sqrt{b}$, the square root of the asymmetry factor. The volume of 'defective' material around individual defects is therefore enhanced over the symmetric geometry. In grazing incidence on the specimen, both the absorption depth and the extinction length become small. Thus, both within and away from the Bragg reflecting range the X-ray wave does not penetrate deep into the crystal. The depth of 'defective' material around the defect becomes comparable with the depth of penetration of the X-ray wave and almost 100% contrast can be obtained from the defect. A browse through the literature will show that this argument is confirmed by experiment and that the asymmetric settings where the reference and the specimen are both in grazing incidence settings give the best defect contrast.

9.5 Synchrotron radiation double-crystal topography

In the laboratory, high spatial resolution, double-crystal topographs can often take several days to expose. The intensity available at synchrotron radiation sources makes the use of this radiation very attractive and exposures at second-generation sources on L4 nuclear emulsions are typically about 30 minutes. This reduces further for the third-generation sources. There are a few features of the synchrotron radiation experiment which differ from that in the laboratory.

1   The very small source size and the large distance of the source from the specimen means that any point on the specimen sees a beam of divergence less than one or two arc seconds. Therefore there is very little difference between the

images in the (+n,−n), (+n,+n) and (+n,±m) settings. Very little dispersion is introduced by mismatching crystals and so highly perfect silicon is recommended as a reference crystal.

2   Because both reference *and* specimen are a long way from the source, the geometric resolution is good for both crystals. Thus, any defects in the reference crystal are imaged in sharp contrast, which is not the case in the laboratory. This provides a second reason for using silicon as a beam conditioner.

9.6   Applications of double-axis topography

Applications of double-crystal topography with synchrotron radiation are dealt with in the next chapter. Here, we confine ourselves to laboratory-based applications.

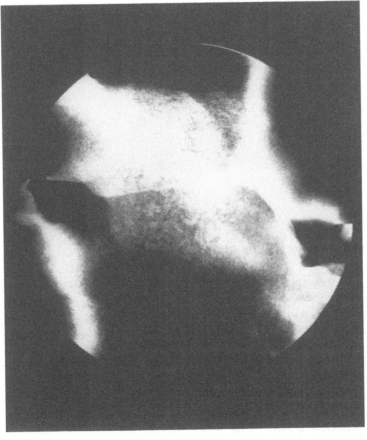

(a)

Figure 9.6   422 reflection double-crystal (asymmetric reflection) reflection topograph of (a) low-angle boundaries in a 3 inch semi-insulating liquid encapsulated (LEC) GaAs wafer; (b) dislocation cell structure in a vertical gradient freeze GaAs wafer. (Courtesy Dr I. C. Bassignana, Nortel Inc, Ottawa)

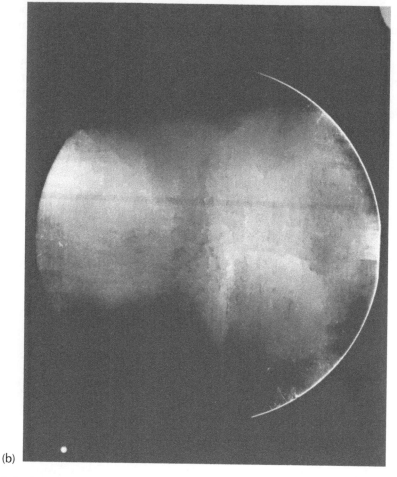

(b)

Figure 9.6 (*cont.*)

9.6.1 *Incoming wafer inspection*

Double-crystal topography has been successfully applied by Bassignana *et al.*[8] to the study of incoming compound semiconductor wafers prior to growth of epitaxial layers (Figure 9.6). Asymmetric reflections are used for both reference and specimen crystals so that the beam expansion is about 20 times and the beam divergence at the specimen is about 1.3 arc seconds. This provides a highly sensitive probe of long range strain. Relatively low resolution, and hence fast, Agfa D7 X-ray film has been used to keep exposures in the region of 20–40 minutes. This was so successful that it was incorporated as part of the routine quality assessment of GaAs wafers purchased by Bell Northern Research and has had significant impact on device yields.[9] Wafers from boules which had a high density of clustered low-angle boundaries were found to be mechanically too fragile to survive all the required steps in device processing. Wafers from boules which showed large included mosaic crystals showed poor or erratic cleaving characteristics, which resulted in the loss of many devices.

216

9.6.2 *Strain contour mapping*

If the specimen wafer is curved, then because of the low angular divergence of the beam, only part of the wafer will be imaged. The incidence angle across the specimen crystal varies with position and a rocking curve is imaged! Figure 9.7 shows a fine example of the image of a curved crystal of GaAs with several AlGaAs epitaxial layers of different composition grown on it. In addition to the peak from the GaAs substrate and that from a thick AlGaAs buffer, a series of interference fringes can be seen across the sample. These match extremely well with the rocking curve taken from a single point under angle scanning.[10]

If there is only the substrate, the image is a single stripe, tracing out the contour of equal effective misorientation. Thus a series of images with different angular separation between reference and specimen crystal provides a 'zebra pattern' of stripes corresponding to different effective misorientation contours. Figure 9.8 shows such a series of contours from a SI LEC GaAs specimen. By rotation of the sample about the surface normal and recording a second series of topographs, the effects of tilts and dilations can be distinguished. Tilts will appear as contour displacements in opposite senses in the two topographs, in which the effect of dilations is the same. Brown *et al.*[11] and Minato *et al.*[12] have given some fine examples of use of zebra stripes for the analysis of long-range strain.

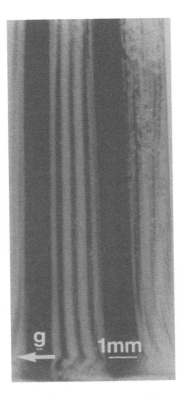

Figure 9.7 Double-crystal topograph taken of a curved GaAs wafer with several AlGaAs epitaxial layers, showing the effective rocking curve spread out in the image[10]

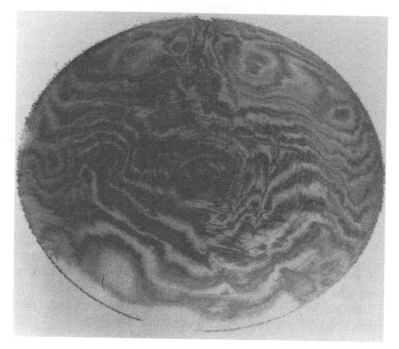

Figure 9.8 Composite image of a series of double-crystal topographs recorded at different incidence angles of a SI LEC GaAs sample. The image is a set of contours of equal effective misorientation. (Courtesy S. J. Barnett)[13]

9.6.3 *Curved crystal reference crystal topography*

The problem of specimen curvature can, as we have seen, be used to advantage, but in many cases it is extremely inconvenient. For low curvature, sharp contour lines do not appear, but only part of the sample is imaged. In order to obtain an image of the whole sample several topographs have to be taken at different incidence angles, multiplying the experiment length by several times (as many times as images are necessary). Jenichen, Köhler and colleagues in Berlin have overcome this by the clever expedient of curving the reference crystal to match the curvature of the sample.[14,15] The dramatic improvement can be seen between the flat reference crystal topograph of Figure 9.9(a) and that when the reference crystal is curved (Figure 9.9(b)).

The minimum radius of curvature is set at a few metres from a combination of geometrical considerations such as primary beam divergence and breakdown of the dynamical diffraction conditions. This does not turn out to be a severe limitation in practice. Highly mismatched or very thick near-matched epitaxial layers turn out to have quite uniform bending which matches well with the mechanical bending deliberately induced on the reference crystal. In the asymmetric geometry, the contrast is high due to the small penetration depth, and excellent quality topographs have been reported. Figure 9.10 shows an example of threading and misfit dislocations in

218

Figure 9.9 Double-crystal topographs with Ge reference crystal, asymmetric 206 reflection, glancing angle 1.7° Cu $K_{\alpha 1}$ radiation. Sample is 2 μm $Al_{0.05}Ga_{0.95}As$, 2 μm $Al_{0.35}Ga_{0.65}As$, 1 μm GaAs on (001) GaAs substrate 206 reflection. (a) Flat reference crystal. (b) Reference crystal curved to 12 m radius of curvature. (Courtesy of Dr R. Köhler)[16]

Figure 9.10 Magnified section of the topograph of Figure 9.9(b) showing threading and misfit dislocations. (Courtesy Dr R. Köhler)[16]

the topograph shown in Figure 9.9(b) but at greater magnification. Note that, when taking topographs of misfit dislocations, it is usually better to use the substrate, rather than the layer reflection. For very thin (typically 20 nm) layers of InGaAs on GaAs this is essential, as the intensity in the layer peak is several orders of magnitude less than that of the substrate and exposure times become prohibitive. For thick layers, the uniformity of the substrate images is usually better. The strain fields of the misfit dislocations are long range and, as the elastic properties of the layer and substrate are very similar, the images appear sharp and without loss of contrast.

9.7 Summary

Double-axis topography can be used at the two extremes of crystal perfection. On the one hand it can be used to probe the long-range lattice strains in very highly perfect crystals. On the other, it can be used to reveal contours of equal misorientation in much less perfect materials. As we will see in Chapter 10, it is widely used at synchrotron radiation sources and with the very large power loading in the white beam of third-generation synchrotron radiation sources. It is rapidly becoming the standard technique of topography for most electronic materials.

References

1. W. L. BOND & J. ANDRUS, Am. Mineralogist, **37**, 622 (1952).
2. U. BONSE & E. KAPPLER, Z. Naturforschung, **13a**, 348 (1958).
3. U. BONSE, in: Direct observation of imperfections in crystals, eds. J. B. NEWKIRK & H. WERNICK (Wiley, New York, 1962), p. 431.
4. S. MERIAM ABDUL GANI, PhD thesis, Durham University (1982).
5. P. RIGLET, M. SAUVAGE, J.-F. PETROFF & Y. EPELBOIN, Phil. Mag. A, **42**, 339 (1980).
6. W. SPIRKL, B. K. TANNER, C. WHITEHOUSE, S. J. BARNETT, A. G. CULLIS, A. D. JOHNSON, A. KEIR, B. USHER, G. F. CLARK, W. HAGSTON, C. R. HOGG & B. LUNN, Phil. Mag. A, **70**, 531 (1994).
7. W. SPIRKL, B. K. TANNER, C. WHITEHOUSE, S. J. BARNETT, A. G. CULLIS, A. D. JOHNSON, A. KEIR, B. USHER, G. F. CLARK, C. R. HOGG, B. LUNN & W. HAGSTON, Phil. Mag. A, **69**, 221 (1994).
8. I. C. BASSIGNANA, D. A. MACQUISTAN & D. A. CLARK, Adv. X-ray Analysis, **34**, 507 (1991).
9. I. C. BASSIGNANA & D. A. MACQUISTAN, 7th Int. Conf. on III–V Semi-insulating materials, Ixtapa Mexico, (1992).
10. S. COCKERTON, PhD thesis, Durham University (1991).
11. G. T. BROWN, M. S. SKOLNICK, G. R. JONES, B. K. TANNER & S. J. BARNETT, in: Semi-insulating III–V compounds, eds. D. C. LOOK & J. S. BLAKEMORE (Shiva, Nantwich, 1984) p. 76.
12. I. MINATO, H. HASHIZUME, H. WATANABE & J. MATSUI, Japan. J. Appl. Phys., **25**, 1485 (1986).
13. S. J. BARNETT, PhD thesis, Durham University (1987).
14. B. JENICHEN, R. KOHLER & W. MOHLING, Phys. Stat. Sol. (a), **89**, 12 (1985).
15. B. JENICHEN, R. KOHLER & W. MOHLING, J. Phys. E: Sci. Inst., **21**, 1062 (1988).
16. R. KÖHLER, Appl. Phys., **A58**, 149 (1994).

Synchrotron Radiation Topography

In this chapter, the application of synchrotron radiation for X-ray topography is reviewed. The intensity and continuous spectrum of synchrotron radiation is particularly important but we see that the time structure and polarisation can also be exploited

10.1 Introduction

Despite becoming gradually accepted as a standard characterisation technique for highly perfect crystals, X-ray topography has a reputation for being slow when performed with conventional X-ray generators in the laboratory. The need for high collimation sets a limit on the elastic strains which can be present in the crystal if high resolution topographs are to be obtained and makes setting up the experiment non-trivial. In 1974 Tuomi et al.[1] showed that synchrotron radiation could be used for X-ray topography with a very simple experimental configuration, which Hart[2] subsequently demonstrated was capable of high-resolution imaging. This led to a major revolution in the type of experiment which could be performed and the nature of the data gathered. The field of synchrotron X-radiation topography has been recently reviewed by the present authors.[3]

10.2 Synchrotron radiation sources

In a conventional tube, X-rays are generated when energetic electrons are stopped by impact with a metal target. The process is inefficient and the thermal properties of solids place an upper limit on the brightness of X-ray beams that can be achieved. [Note that *brightness* is *flux per unit solid angle* and while flux may always be increased by increasing the area of target used for electron impact, this does not increase the brightness. As we have seen, in a topographic experiment the spatial resolution deteriorates as the source size increases.]

When electrons are confined to a circular orbit in a storage ring by a magnetic field, electromagnetic radiation is emitted owing to their acceleration towards the centre of the circle. At relativistic velocities, the emission is strongly peaked in the forward direction, tangential to the electron orbit (Figure 10.1(a)). Perpendicular to the orbit plane, the emission angle θ is given approximately by $\theta = m_0 c^2 / E$, where m_0 is the electron rest mass, c is the velocity of light and E is the electron energy. At

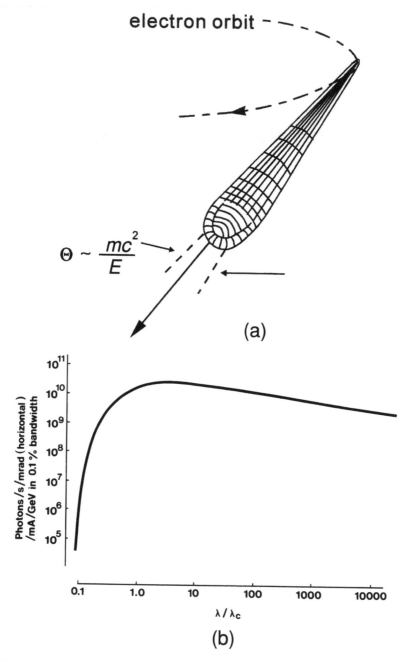

Figure 10.1 (a) Distribution of emitted radiation from an electron travelling at relativistic speeds in a circular orbit. (b) The universal, vertically integrated, synchrotron radiation spectrum as a function of reduced wavelength. The axes are calibrated for the SRS at Daresbury Laboratory

5 GeV, we find that the vertical emission angle is naturally collimated to 0.1 mrad. In the orbit plane, the emission is integrated over a range of natural collimation angles as the electron sweeps round the ring and in practice the beamline acceptance angle limits the divergence in the horizontal plane.

The relativistic transformation from the electron frame to the laboratory frame results in a spectrum which is a continuum extending from the radiofrequency to the X-ray region (Figure 10.1(b)). This is a 'universal' curve scaled to a parameter called the critical wavelength λ_c. The low wavelength fall-off is determined by the energy of the electron beam and the radius of curvature at the point of emission. As there are no target materials to melt, the brightness achievable from synchrotron radiation sources may be 10^3 to 10^8 times higher than those obtained from standard X-ray generators.

In a storage ring designed for production of synchrotron radiation (Figure 10.2) the orbit is not circular but consists of dipole bending magnets separated by straight sections. The beam is focused by quadrupole magnets, forming a complex 'lattice' of magnets. After initial acceleration, to a few tens of MeV in a linear accelerator, electrons (or sometimes positrons) are accelerated to the full energy of 2–10 GeV by a booster synchrotron, either alone or in combination with acceleration in the storage ring itself. When the full energy is reached the magnetic field is held steady (apart from dynamic control adjustments to keep a stable orbit), the electron energy is maintained once or more per revolution by a radiofrequency accelerating cavity. The electron beam current decays exponentially by collision with residual gas molecules in the ultra-high vacuum and lifetimes for decay to 1/e of the initial value may be as long as 70 hours.

Beamlines are constructed to extract the beam from tangent points of the bending magnets or from insertion devices within the straight sections. In the latter, the electron beam is deflected in a small radius of curvature by a succession of dipole magnets of alternating sense. There are of two types: wigglers, in which the ampli-

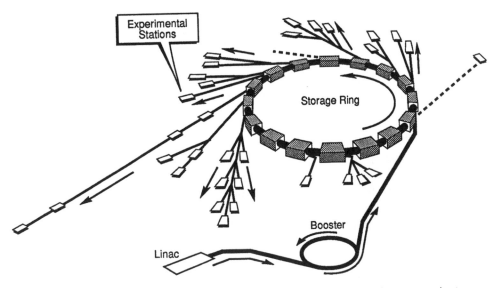

Figure 10.2 Schematic diagram of an electron storage ring showing the magnet lattice and the form of the orbit

223

tude of oscillation is large with respect to the cone of emitted radiation, and undulators in which it is small. Recent devices all exploit permanent magnets of neodymium iron boron to produce high fields in small gaps which can be varied according to the experiment being performed. The spectrum of radiation produced by a wiggler is similar to that from a bending magnet, but with the wavelength shifted to a shorter value by virtue of the higher magnetic fields that are used. Use of a multipole device enables the power in the beam to be increased proportionally to the number of periods in the device. In the third-generation sources such as the ESRF at Grenoble, in France, the APS at Argonne, in the USA, and SPring-8 near Kobe in Japan, most of the beamlines are served by such insertion devices where the beam is deflected and then resumes its original orbit.

The spectrum from an undulator is very different, and numerous peaks result from interference effects within the undulator. When the electron acceleration is confined to the orbit plane and the emission angle very low, the radiation is strongly elliptically polarised and, in the orbit plane itself, it is to within a few per cent linearly polarised. Use of a sequence of permanent magnets with magnetisation arranged in a spiral sequence enables circularly polarised radiation to be extracted from such a helical undulator and this radiation is particularly important for magnetic studies.

The electrons are bunched by the action of the RF accelerating field and therefore the emission occurs in sharp pulses, only tens of picoseconds in width, with frequency corresponding to the bunch spacing. For maximum beam current many bunches are used, giving a pulse frequency of typically 500 MHz. When only one bunch is injected into the ring, the period drops to a few megahertz, while the pulse width remains at a few picoseconds. This feature has been used extensively for fluorescence lifetime measurements and may also be exploited in stroboscopic topography.

10.3 Source requirements for synchrotron X-ray topography

In the laboratory, X-ray topography is usually performed with characteristic radiation and it is straightforward to show that the exposure time for a section topograph scales with PS_x^{-2} and for a traverse topograph as PS_x^{-1}, where P is the X-ray tube power and S_x is the source dimension in the incidence plane. The continuous nature of the synchrotron spectrum, however, results in a very different dependence on the source power and dimensions. Intensity dI reaching the crystal due to area dX, dY of the source is

$$dI = CP(\lambda)\delta\lambda \, dX \, dY / (S_x S_y D^2)$$

(10.1)

where C is a constant, P is the power, S_x is the horizontal size of the source, S_y is the vertical size of the source, and D is the source to specimen distance. We assume that the specimen is smaller than the cone of radiation. We also assume that the angular width of the source seen by a point on the specimen $\Delta\theta_s$ is larger than the intrinsic range of the Bragg reflection $\delta\omega$. (This is valid for first-generation sources and for high energy radiation weak Bragg reflections.)

$$I = CP(\lambda)\delta\lambda \int_0^{D\delta\omega} dX \int_0^{S_y} dY / (S_x S_y D^2)$$

(10.2)

where $\delta\omega = \text{constant}.|C|\lambda^2 b^{1/2} F_h / \sin 2\theta_B$. Thus

$$I = CP(\lambda)\delta\omega\delta\lambda/(S_xD) \qquad (10.3)$$

Consideration of the duMond diagram (Figure 10.3(a)) shows that

$$\delta\lambda = \lambda \cot\theta \, \Delta\theta_s \qquad (10.4)$$

where $\Delta\theta_s = S_x/D$, the angle subtended by the source. The intensity I_h is given by

$$I_h = A|C|\lambda^3 P(\lambda)F_h b^{1/2}/D^2 \sin^2\theta_B \qquad (10.5)$$

where $b = \gamma_0/|\gamma_h|$, $C = 1$ or $\cos 2\theta_B$, F_h is the structure factor, A is a constant and θ_B is the Bragg angle. D is the specimen to source distance, $P(\lambda)$ is the source power per unit wavelength interval, $\gamma_0 = \mathbf{K}_0 \cdot \mathbf{n}$ and $\gamma_h = \mathbf{K}_h \cdot \mathbf{n}$, where $\mathbf{n}$ is a unit vector parallel to the inward normal to the specimen surface. $\mathbf{K}_h$ and $\mathbf{K}_0$ are the diffracted and incident wavevectors.

Now let the source angle $\Delta\theta_s$ be much less than the reflecting range of the crystal $\delta\omega$. This corresponds to low order, long wavelength reflections at third-generation sources. Then

$$I = CP(\lambda)\delta\lambda \int_0^{S_x} dX \int_0^{S_y} dY/(S_xS_yD^2) = CP(\lambda)\delta\lambda/D^2 \qquad (10.6)$$

Now, as seen in Figure (10.3(b))

$$\delta\lambda = \lambda \cot\theta_B \, \delta\omega \qquad (10.7)$$

and again we have

$$I_h = A|C|\lambda^3 P(\lambda)F_h b^{1/2}/D^2 \sin^2\theta_B \qquad (10.8)$$

There is thus no complicating factor involving source dimensions on going from the first to the second criterion used above. Importantly, the intensity is not dependent on the source size but rather on total source power. However, the resolution R_x in a topograph is given approximately by

$$R_x = dS_x/D \qquad (10.9)$$

where d is the source to detector distance. There is thus gain in intensity with increased brilliance if D is reduced, or gain in resolution if D remains constant.

The limit on reduction of D is the beam height at the specimen. There is a requirement to image an area of about 10mm by 10mm to half intensity across the Gaussian beam intensity profile. Then as the intrinsic divergence $m_0c^2/E = 8 \times 10^{-5}$ for a 6GeV machine, this gives $D = 100$m for a typical topography beamline. The resolution is excellent and $R = 0.5\,\mu$m for specimen to detector distance of 10cm with a 0.5mm half-width source size.

Multipole wigglers clearly have major advantages in the enhanced power available for topographic experiments but heat load is a serious problem. Owing to the non-uniform beam, which has a Gaussian profile in intensity there is non-uniform heating. The crystal is locally bent and, when the angular deviation Δ due to the lattice strain exceeds the rocking curve width $\delta\theta$, a strong band of contrast is observed. Barrett et al.[4] have shown that the maximum angular variation Δ from heating should be

$$\Delta = \sqrt{\pi}\,\alpha P_{abs}\,\sigma/2K \qquad (10.10)$$

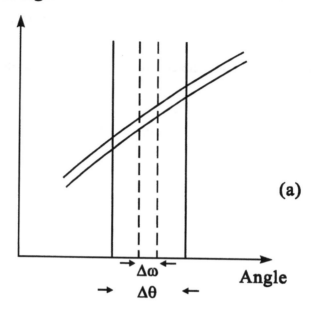

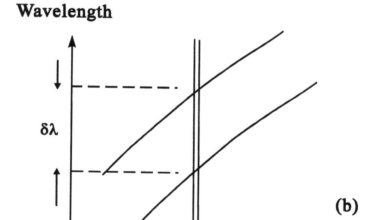

Figure 10.3 (a) duMond diagram corresponding to a synchrotron radiation source where the angular width of the source is larger than the perfect crystal reflecting range, (b) duMond diagram where the reverse is true

where α is the linear thermal expansion coefficient, K is the thermal conductivity, σ is the rms width of beam profile and P_{abs} is the power absorbed. Ultra-short exposure times, on microsecond timescales, are unlikely to be achieved for white beam topography owing to this heat loading problem. It is particularly a problem for low conductivity insulators such as quartz and HgI_2. With high levels of bending, direct image contrast falls as the 'perfect' crystal background rises and the defect images become invisible. Solutions to the problem can be to use a chopper in front of the specimen to throw away up to 90% of the intensity and to filter out short wavelengths with an absorber of aluminium between 0.5 to 1.5 mm thick. A further reduction can be achieved by 'cooling' the beam by passage through about 50 mm of water. Inclusion of a solution containing the element under investigation reduces the intensity around the absorption edge disproportionately and works very well.

A solution to the heat load problem is to use a highbandwidth monochromator to reject most of the incoming wavelengths. However, mirrors are at present not sufficiently uniform to be usable and the figuring error on the commercially available synthetic multilayers is too great for topographic use. Revol et al.[5] have demonstrated the use of a vibrating silicon crystal as a broadband monochromator and in any case a variable gap wiggler is essential. For long wavelength work the wiggler gap should be increased.

The spectrum of an undulator which, owing to interference of the coherent X-radiation emitted by the electron on successive excursions within the magnet, consists of a series of peaks, looks at first sight to be ideal for synchrotron X-ray topography. The energy width of the undulator peaks is large compared with the perfect crystal reflecting range and quasi-white beam topography could be performed without the heat load problems of a multipole wiggler. In addition, the coherence of the undulator radiation results in it being many orders of magnitude brighter than that from a multipole wiggler. The beam is of low divergence but at 400 m is sufficiently wide to meet the beam-size criterion given above. However, the spectrum from an undulator varies rather fast as a function of angle and this means that the beam arriving at the sample has a different spectral characteristic at any point. As seen from Figure 10.4(a), the crystal picks out a narrow band of energy to diffract and, for a given setting, the intensity in the diffracted beam falls much more rapidly than expected (Figure 10.4(b)). Unless a beamline of 1 km length is used, undulators do not seem to be ideally suited to X-ray topography.

10.4 Applicability of synchrotron radiation to X-ray topography

There are five features of synchrotron radiation which make it both unique and worth the inconvenience of travelling to a distant place to conduct experiments.

10.4.1 *Intensity*

In order to take high resolution X-ray topographs, slow, very small grain size photographic emulsions must be used. Ilford L4 nuclear emulsions give the best results, and exposure times on conventional generators range from hours or days. Use of synchrotron radiation reduces exposure times on these emulsions to a matter

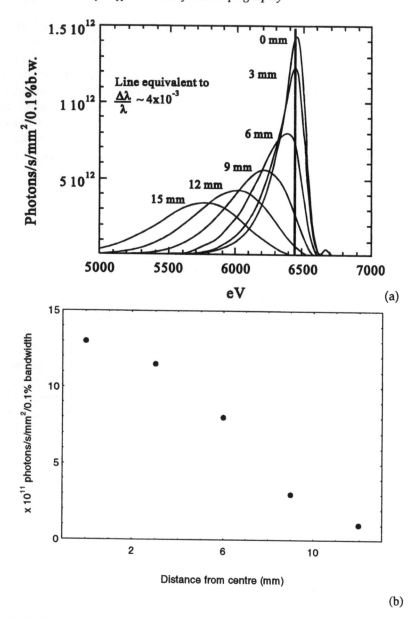

Figure 10.4 (a) Spectrum, at 400 m distance from an ESRF undulator, for different positions across the image with band of wavelengths diffracted by a typical crystal. (b) Diffracted intensity as a function of position across the image

of seconds or less. Direct imaging TV detectors, which are severely limited in resolution from quantum noise with conventional generators have enough photons to give good quality images in real time. Depending on the area imaged, the flux available at the detector and the type of device, resolutions between 40 and 8 micrometres have been achieved. Figure 10.5 shows an example of low-angle boundaries and magnetic domains in a crystal of iron–silicon alloy imaged in real

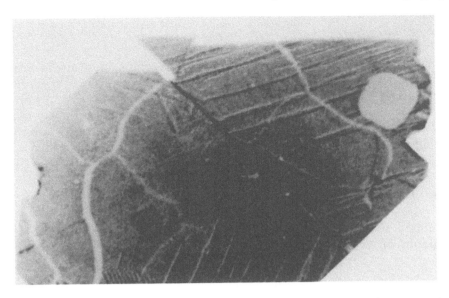

Figure 10.5 Real-time image, with no computer processing, of magnetic domains and low-angle boundaries in iron–silicon taken using the Bede Scientific HI-RES detector at Daresbury Laboratory. The white radiation topograph was taken with X-rays of fundamental wavelength 1 Å

time with an X-ray sensitive TV camera at the Daresbury SRS. *Dynamic* experiments, in which changes in the material are studied as a function of variables, such as time, temperature, stress or electric or magnetic field, may thus be performed either in real-time at moderate resolution or in a rapid step-by-step mode at high resolution. *Survey* experiments in which a very large number of samples are studied to provide statistically meaningful data may also be undertaken.

10.4.2 *Continuous spectrum*

A storage ring designed for synchrotron radiation emits radiation over a very broad spectrum (Figure 10.1(b)) which extends down to 0.1 Å at the third-generation sources. The long wavelength cutoff is usually set at about 2.5 Å from the absorption of the beryllium window. This wide wavelength range allows one to optimise independently geometry, absorption conditions and strain sensitivity. The continuous spectrum permits high resolution Laue topographs to be taken which are extremely easy to set up and are applicable to a wide range of materials. In addition, it is possible exploit the dispersion near absorption edges.

10.4.3 *Collimation*

Because of the low divergence, topography experiments may be conducted at a great distance (50–1000 m) from the source. This provides a wide beam at the specimen and large areas may be illuminated without the need for scanning.

10.4.4 Polarisation

The high degree of X-ray polarisation in the electron orbit plane provides means of controlling both the signal/noise ratio and the penetration of the X-rays into the specimen. Depending on whether the incidence plane is chosen vertically or horizontally, sigma or pi polarisation may be selected. The strain sensitivity and the extinction distance can thus be varied while the normal photoelectric absorption conditions remain identical.

10.4.5 Time structure

The pulsed time structure, typically at megahertz rates, permits stroboscopic topography of very rapid cyclic phenomena such as the propagation of acoustic waves.

10.5 Techniques of synchrotron radiation topography

10.5.1 White radiation topography

In a conventional Laue photograph, when 'white' (that is continuous) radiation is allowed to fall on a single crystal, each set of planes selects a particular small slice of the spectrum appropriate to the angle between the Bragg planes and the incident beam to satisfy the Bragg condition. Many diffracted beams are therefore observed in a pattern characteristic of the symmetry of the crystal structure and the orientation of the crystal with respect to the beam. From this, Guinier and Tennevin[6] and Schulz[7] had devised early continuous radiation laboratory-based topographic techniques but the resolution was poor. Not until synchrotron radiation became available did the power of these simple techniques become apparent. When the source is a large distance from the specimen, and the width of the beam at the sample large, each large Laue 'spot' becomes a high resolution map of the scattering from different points across the crystal. Typical numbers inserted into equation (10.9) show that a high spatial resolution can be preserved in these diffraction topographs for most synchrotron radiation sources. Figure 10.6 shows schematically how the Laue topographs arise.

Around defects, the scattering power differs from that in the perfect crystal because X-rays which do not satisfy the Bragg condition in the perfect crystal may be diffracted in the deformed region around the defect. Just as in the Lang projection topograph, these regions behave as small crystals which diffract kinematically and the net result is an increase in the intensity over that from the perfect crystal.

For a given reflection, we can deduce from equations (10.5) and (10.8) that:

- the polarisation factor, which is 1 for vertical dispersion and $|\cos 2\theta_B|$ for horizontal dispersion, has little effect at short wavelengths and a horizontal dispersion geometry can be used with little intensity loss
- the intensity falls off with decreasing wavelength due both to λ and $P(\lambda)$

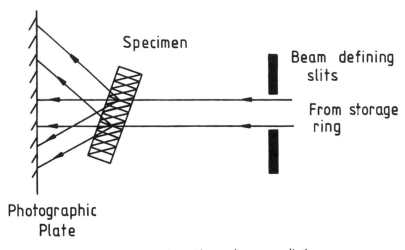

Figure 10.6 White radiation topography with synchrotron radiation

Defect contrast is similar to that in Lang topographs[8] but with differences due to:

- the presence of higher harmonics diffracted simultaneously
- the divergence of the beams between crystal and film, and
- the continuous nature of the radiation spectrum

The contribution of harmonics is not normally large at first- and second-generation synchrotron sources, as the F_{hkl} value falls rapidly with increasing diffraction vector. A major exception is where weak quasi-forbidden reflections such as 002 are being used in the III–V compounds. Here the 004 harmonic at $\lambda/2$ is extremely important as $F_{002}(\lambda) \ll F_{004}(\lambda/2)$. However, for third-generation sources operating at high energy, the critical wavelength which determines the point at which the spectrum rolls off is very short. For a fundamental at about 1.5 Å, the incident beam flux at harmonics up to fifth order will be very high. In the (Laue) transmission geometry, the low orders are attenuated by photoelectric absorption more than the higher harmonics and this compensates for the loss in intensity in the higher harmonics predicted by equations (10.5) and (10.8). Thus, at the ESRF for example, the harmonic contribution to the image is extremely important, whereas at a second-generation source such as Daresbury, the problem is minimal. As a result the image width as a function of wavelength may not be a monotonic function and in some instances increases with increasing wavelength (Figure 10.7). This is contrary to the prediction in Chapter 8 that the width of dislocation images increases with decreasing wavelength.[9,10]

A second difference in contrast arises due to the large specimen-to-plate distances which are used in order to separate the various Laue topographs. Then the angular deviation of the beams diffracted from around the defect[8] becomes significant. The beams diffracted kinematically from either side of an edge dislocation (Figure 10.8) diverge or converge according to the effective misorientation around the defect. Dislocations with edge components often show this bimodal

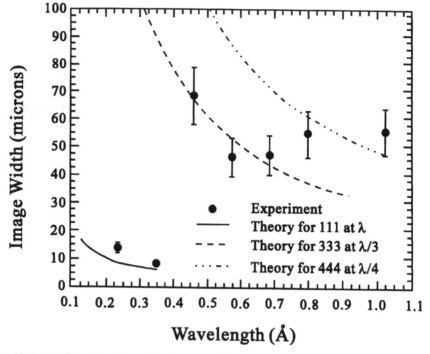

Figure 10.7 Width of a dislocation image under conditions where harmonics are strong as at the ESRF. (Courtesy F. Zontone)

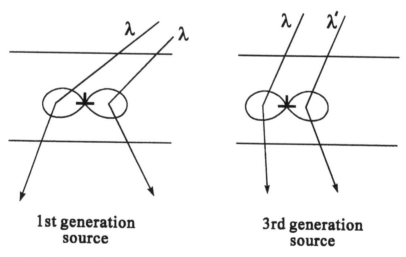

1st generation
source

3rd generation
source

Figure 10.8 Mechanism for formation of the bimodal image seen in many synchrotron radiation topographs

contrast which is only seen in the laboratory for high values of the product **g.b**. The separation of the lobes increases linearly with distance of the detector from the specimen.

The third difference arises from the continuous nature of the radiation giving rise to orientation contrast only at the boundary between misoriented regions. Here the beams overlap or diverge depending on the sense of the misorientation with respect to the diffraction vector (Figure 10.9(a)). Best contrast occurs when the beams diverge, where there is a total lack of intensity associated with the boundary region. An example is given in Figure 10.9(b) of sub-grain boundaries in $Tb_{0.27}Dy_{0.73}Fe_2$. The magnetic domain contrast in Figure 10.9 arises because the magnetostriction leads to a tetragon al distortion of the lattice. At a 90° boundary

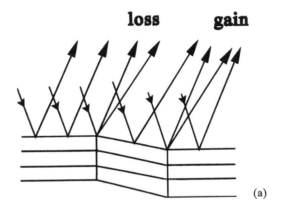

(a)

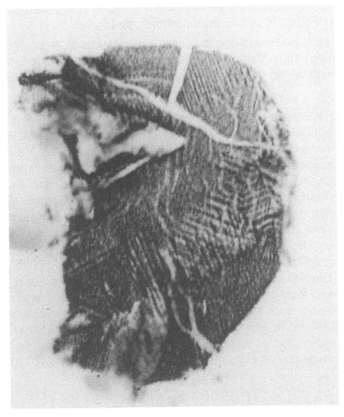

(b)

Figure 10.9 (a) Schematic diagram of the deformation of orientation contrast images in white radiation topography. (b) Low-angle boundaries in $Tb_{0.27}Dy_{0.73}Fe_2$ showing divergent (white) contrast

233

the lattice is coherent along a (110) plane, but the {100} planes are tilted on crossing the boundary. This misorientation results in orientation contrast similar to that of a sub-grain boundary.

White radiation topography has the following special features:

1 Short exposures (from milliseconds up to minutes, depending on the source and the specimen).

2 Many reflections recorded at once on one plate.

3 The ability to image defects in crystals which contain long-range elastic strains as each point on the specimen selects its own appropriate wavelength band for diffraction.

4 The ability to obtain overall strain maps of specimens that are too distorted for individual defects to be distinguished.

5 A simple experimental arrangement that is well suited for experimental stages for *in situ* control of experimental variables.

It is therefore particularly well-adapted to the following types of study:

1 *Dynamic experiments* involving the study of changes in materials in real time, as a function for example of temperature, stress or magnetic and electric fields.

2 *Distorted crystals* where the lattice perfection is poor but information is required of the distribution of lattice strain through the crystal.

3 *Survey experiments* involving the rapid examination of large numbers of samples, or large numbers of reflections in order to obtain detailed statistical information relating to growth technique or detailed three-dimensional strain analysis.

Most of the early work on synchrotron radiation topography was performed with very simple goniometers to provide basic specimen orientation. The specimen is placed in the beam and a plate or film used to record the image. However, in order to set up a reflection at an arbitrary wavelength one needs at least two axes on each of the specimen and detector, and to control polarisation one must be able to rotate, albeit perhaps indirectly, around the incident beam. To perform these functions quickly, for alignment or to perform dynamic experiments, remote computer control is now essential, integrated with data collection. An excellent example is the 'Orient Express' program at the ESRF. A computer program to calculate the Laue pattern for an arbitrary specimen and detector orientation,[11] or to index a Laue pattern from the position of several images is an important adjunct to the station control software. Real-time imaging with acceptable dynamic range and resolution is now a possibility with the high fluxes at the most recent synchrotron radiation sources. Without doubt, however, the major limitation in the exploitation of X-ray topography to dynamic, *in situ* experiments is in the performance of current detectors. The performance of a quantum limited detector is governed by the Rose–de Vries law,

$$C^2 \delta^2 I \tau \xi / k^2 = 1 \tag{10.11}$$

where C is the contrast of the defect, δ is the resolution or pixel size, I is the intensity, τ is the exposure time, ξ is the detector quantum detection efficiency and k is the system signal-to-noise ratio. While improvements in spatial resolution are clearly desirable, it is economically unrealistic to do other than exploit technology developed for alternative commercial purposes. Rather slow progress has been made in this area, which remains of critical importance to the effective use of the third-generation sources. A micron resolution CCD detector with moderate read-out rate is now operating at the ESRF.

The main reason for requiring rapid acquisition of different images is for detailed strain analysis. We have seen in Chapter 8 that the Burgers vector of dislocations can be found by noting that there is zero contrast for a screw dislocation when $\mathbf{g}.\mathbf{b} = 0$ and for an edge dislocation when, in addition, $\mathbf{g}.\mathbf{b} \times \mathbf{l} = 0$, where $\mathbf{g}$ is the reciprocal lattice vector of the reflection, $\mathbf{b}$ is the Burgers vector and $\mathbf{l}$ the line direction vector of the dislocation. Selection of sufficient reflections will then permit this determination, though in practice there may be difficulties caused by anisotropic elastic behaviour and by surface relaxation. With white radiation topography, several reflections can be recorded simultaneously, with enormous time-saving over laboratory investigations. In Figure 10.10, many reflections are recorded at once, providing enough information to characterise the twin boundaries. With a conventional generator this would have taken about a week.

The complete strain tensor in a specimen can be obtained by sufficient topographic images, if a narrow slit is used to define a direction, or very sensitively, an internal marker is obtained by choosing the wavelength range on the image so that it crosses an absorption edge. The methodology has been worked out and applied to strains in metals by Stock and colleagues.[12]

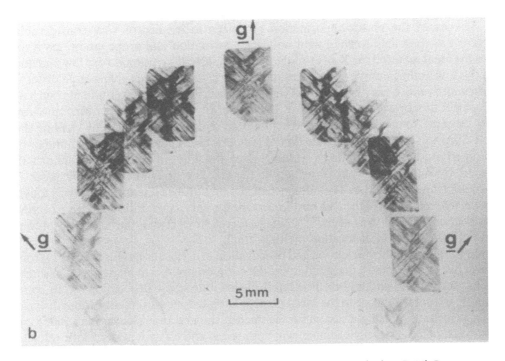

Figure 10.10 Laue topographs taken simultaneously from a crystal of NaBaNbO$_3$

10.5.2 Example applications of white radiation topography

In our review article cited above,[3] we have given a detailed review of the work in the field up to 1994. Therefore in this section, we will give only a few examples to indicate the power and range of application of the technique.

Relatively little work has been done using section topography with synchrotron radiation. Machado and colleagues[13] have used it to study thick highly perfect diamonds where the flexibility of choosing the Bragg angle for a particular reflection over a considerable range permits the experimenter to obtain several different views of the crystal interior whilst still retaining the same diffraction vector. Halfpenny has also exploited this at Daresbury Laboratory for *in situ* solution growth studies.

Tuomi *et al.*[14] used the Stanford SSRL source for section topography studies of processing induced defects in semiconductors. In a study of intrinsic and extrinsic gettering of silicon they observed the loss of the 'perfect crystal' interference fringe pattern in the section topograph after initial oxidation and its recovery after the well drive cycle at 1150°C for 18 hours. This indicated that the microdefects formed during the initial oxidation were dissolved during the drive cycle. After field oxidation at 950°C for 10 hours, a very high number of microdefects were created which totally destroyed the interference (Kato) fringes. In the section topographs the denuded zone was clearly visible and the technique provides a non-destructive means of measuring the width of this region. Correlations were found between gettering defects revealed by synchrotron radiation section topography and device yield. The best device yield was observed from the middle sections of wafers, with the rear surface covered by $2\,\mu$m of polysilicon. These discernible precipitates or stacking faults do not develop after the complete thermal cycle.

Studies of the plastic deformation processes in ice are of very considerable interest as they help to shed light on glacier formation and propagation. Ice is an almost ideal material for X-ray topography in that its absorption is very low and due to the low atomic number, the structure factors F_h are low, thus giving large dislocation widths which may be recorded very satisfactorily on low resolution film such as Kodak R film or even a TV detector. Hondoh *et al.*[15] used a direct conversion X-ray TV camera[16] to measure directly the climb rate of dislocations, thereby yielding the self-diffusion coefficient as a function of temperature. There have been a number of detailed studies, by Hondoh's group[15,17,18] at the Photon Factory at Tsukuba in Japan, by Whitworth[19-22] at Daresbury in the UK and by Dudley and colleagues at Brookhaven in the USA. The influence of slip on non-basal planes in ice was found to be unexpectedly high, and on these planes a very large ratio of edge to screw dislocation velocity was seen.[17,20,21] Image forces near crystal surfaces also played an important role in dislocation multiplication.

In the study of the nucleation and propagation of dislocations (Figure 10.11) from scratches in a geometry designed to produce slip on non-basal planes, edge dislocations were found to glide on non-basal planes, but screw dislocations were completely immobile except in the basal plane.[21] Detailed measurements of the dislocation velocities[17,22] using stress pulse techniques showed that dislocations glided on the basal plane as straight segments in both screw and 60° orientation with velocity proportional to stress. The formation of straight glide segments in the basal plane is evidence that dislocations glide by propagation of kinks across a Peierls barrier. However, activation energies determined for the two types show marked order of

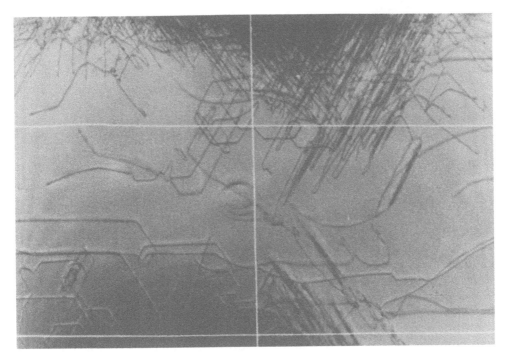

Figure 10.11 Dislocation multiplication in ice during plastic deformation. (Courtesy Dr R. Whitworth)

magnitude disagreement with current theoretical models. The activation energy data for non-basal dislocations suggests that proton disorder may limit dislocation mobility, though there is still a disagreement between the predicted and observed magnitudes for the dislocation velocities.

The Stonybrook group under Dudley has studied the behaviour of ice bicrystals and has shown that under certain conditions, grain boundaries can act as sources of dislocations. Grain boundary facets have been shown to act as dislocation nucleation sites and grain boundaries themselves have been observed to act as barriers to dislocation motion.

In the 1970s there was considerable interest in antiferromagnetic domains. As there is no net magnetization in such domains, decoration and Lorentz microscopy techniques cannot be used to reveal them. Polarised light can only be used to reveal the birefringence associated with the lowering of the symmetry in the antiferromagnetic phase if the magnetostriction is large. X-ray topography is much more sensitive to the magnetostrictive distortion and can be used to reveal the anti-ferromagnetic domain structures[23] in materials such as $KNiF_3$ where the magnetostriction is of the order of parts in 10^5. Quasi-static SR studies of $KNiF_3$ and $KCoF_3$[24] in moderate magnetic fields generated by an iron yoke electromagnet, and at temperatures down to 4.2 K, provided the first experimental confirmation of Neel's theoretical predictions concerning the behaviour of cubic antiferromagnets in applied fields. Domain boundaries move in such a way as to increase the volume of material where the sub-lattice magnetisation is normal to the applied field, thus reducing the free energy of the system. The experimentally observed displacement as a function of field was in very good agreement with the predictions of Neel's theory. Figure 10.12 shows an example of a series of topographs taken at temperature of 4.2 K as the magnetic field is increased.

Recrystallisation and grain growth after deformation has been studied by *in situ* SXRT experiments by Gastaldi, Jourdan and co-workers for over a decade.[25] X-ray topography will not reveal anything until new grains are more than a few micrometres in size, and so it is not useful for nucleation studies, but it is excellent for overall distribution measurement of new grains, correlation with other microstructure, kinetic studies and determination of defect structures after grain growth. *In situ* kinetic studies, for example, enable the range of grain growth kinetics to be extended to some 10–100 times higher velocities than is possible with interrupted annealing experiments.

The work of Gastaldi and Jordan has provided important insights into the interaction of moving grain boundaries and formation of defects when they collide. Their early studies were performed only under high vacuum but subsequent redesign of the furnace has permitted work to be achieved under ultra-high vacuum conditions. *In situ* measurements of the grain boundary displacements at various temperatures have been made[26] and it has been shown how grains which nucleate at pre-positioned surface indentations and subsequently grow into the pre-strained matrix, transform progressively from random to faceted configurations. Analysis of migration rates as a function of temperature yielded activation energies a factor of five

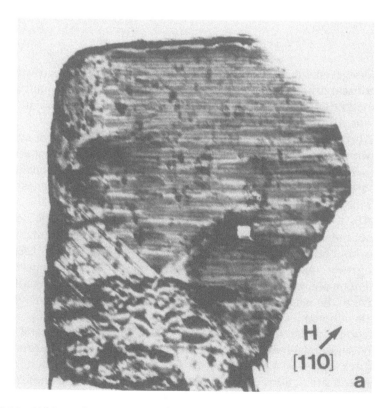

Figure 10.12 White radiation topographs of 90° antiferromagnetic domain walls in KNiF$_3$ under conditions of increasing magnetic field. (a) 0.27 T, (b) 0.46 T (c) 0.64 T

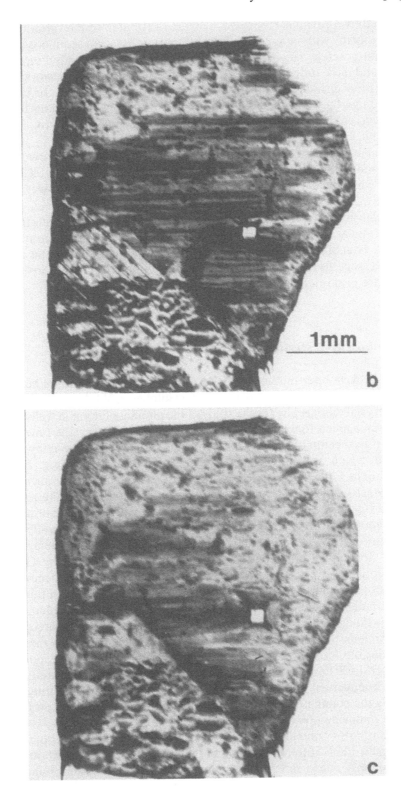

Figure 10.12 (*cont.*)

greater than those previously reported, a feature attributed to the variation in grain boundary mobility with grain boundary inclination as a consequence of growth selection. Matrix screw dislocations have been seen to be left behind moving grain boundaries and these defects are incorporated in the dislocation network produced by reaction with dislocations originating from the grain nucleus. In contrast to previous findings, they concluded that the density of matrix dislocations left behind did not depend on the migration rate,[25] leading to the proposal of a new dislocation generation mechanism.

The group has more recently undertaken studies of the melt-growth of alloy systems in which the liquid–solid interface is not planar. Figure 10.13 shows a sequence of white beam topographs taken during melt growth of an AlCu alloy. As the interface advances, the material which solidifies is relatively perfect, with a low dislocation density. However, at the growth rate here (4.1 μm/s) the interface is not stable and cellular growth results. The cell spacing is a continuous function of growth rate. Eventually molten liquid becomes entrapped behind the advancing interface and from these regions, dislocations are generated. The key insight is that the defects are generated a long way behind the advancing interface.[27]

10.5.3 Stroboscopic studies

Standard optical chopper methods may be applied to stroboscopic studies of cyclic dynamic phenomena up to frequencies of a few kilohertz. While this has been used in multiple crystal topography for the study of the pinning of domain walls in Fe–Si[28] it has not been applied to white radiation topography. However, the time structure of the synchrotron beam itself may be used in high frequency stroboscopy. In single-bunch mode, where only one bunch of electrons orbits in the ring, X-rays are emitted in bursts typically picoseconds in width, spaced by a few microseconds. Using this MHz signal taken from the beam position monitors in the ring, a high frequency device can be excited phase locked to the synchrotron radiation emission. This high frequency stroboscope was first utilised in the study of travelling surface acoustic waves in SAW devices (intermediate frequency filters and delay lines), by Whatmore, Goddard, Tanner and Clark[29] and bulk wave devices by Graeff and Gluer.[30–32] Until these experiments were undertaken, it had only generally been possible to observe the time-averaged displacement. Loss of transmitted acoustic power in surface acoustic wave (SAW) delay lines had been a problem for some time and was thought to be associated with lattice defects. The stroboscopic topography experiments, together with other evidence, showed conclusively that the main defect responsible for the SAW scattering was a region of reverse polarisation, known as a 'rille'. These reverse ferroelectric domains were associated with low-angle grain boundaries, their interaction with the boundary preventing their removal during the poling process. The defects were shown to initiate new bulk wave modes, thus removing power from the surface wave.

An illustration of travelling surface waves revealed by stroboscopic SXRT is shown in Figure 10.14. Here, interaction of the wave with a sub-grain boundary can be clearly seen. A phase shift is introduced and, at the sub-grain boundary, new modes are excited. In particular one may identify the second- and fourth-order harmonics. Thus, although the defect does not absorb the acoustic power, it results

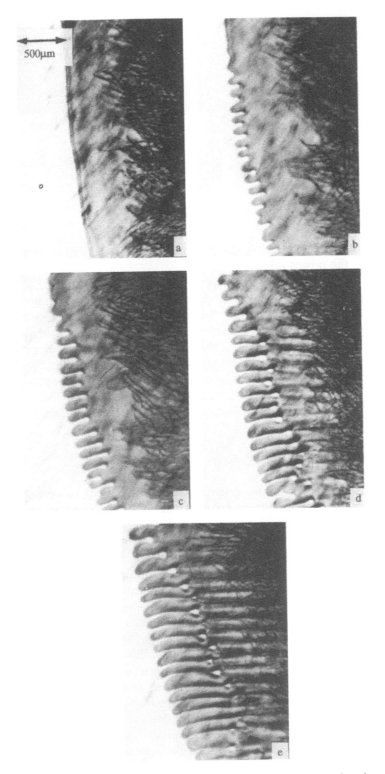

Figure 10.13 *In situ* synchrotron radiation X-ray topographs of the interface between melt and solid in an AlCu alloy. (Courtesy G. Grange and J. Gastaldi)

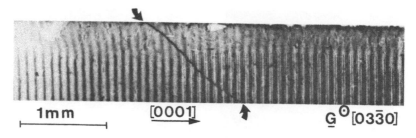

Figure 10.14 Stroboscopic topograph of travelling surface acoustic waves in a LiNbO$_3$ IF filter

in a redistribution of power into unwanted modes of excitation. From an engineering viewpoint, the power is lost.

Bulk wave devices have different tolerances and recently Capelle, Zarka and co-workers have studied bulk waves in quartz resonators and used stroboscopy to identify unwanted modes associated with defects.[33] They have also performed true section topography in stroboscopic mode to identify if the interaction between a dislocation and the acoustic wave could be described by simple linear piezoelectric theory. Using simulation of the section topographs to analyse the data, they concluded that a non-linear interaction was present *near* to the dislocation line, linear theory working satisfactorily in the region far from the defect. Etch channels appeared to have more influence on the acoustic wave than individual dislocations.

10.5.4 *Multiple crystal topography*

The white beam results in a high background and insensitivity to long-range strains. Both problems are overcome by use of a crystal before the specimen, which, owing to the small intrinsic angular divergence of the synchrotron radiation beam, acts as a monochromator. This is illustrated in the duMond diagram of Figure 10.3(b). The area is phase space is small and we see immediately that the reflecting range of the monochromator sets the wavelength dispersion, while the angular divergence is set by the source characteristics. Here, unlike the laboratory case, we can think of the first crystal as a genuine monochromator. As discussed in Chapter 9, the high collimation, and hence small dispersion, results in the $(+n,-m)$ and $(+n,+m)$ settings giving almost identical results. With synchrotron radiation, it is often convenient to use a channel-cut crystal or a two-crystal Golovshenko–Graeff device, in what is strictly the double-crystal geometry. The pair of reflections throws the beam into the forward direction and greatly simplifies the experiment.

If the (flat) specimen is rotated in the dispersion plane with respect to the beam conditioner the rocking curve is extremely narrow and is the correlation function of the two perfect crystal reflecting ranges. This is typically a few seconds of arc. If the specimen *or* the beam conditioner contains a region misoriented by more than this amount, no diffracted intensity reaches the detector from that region. Thus, a photographic plate will record a topograph with very high strain sensitivity. The contrast in double-crystal topographs is mainly orientation contrast and has been treated by Bonse[34] (see Chapter 9). There are several differences between double-crystal topography in the laboratory and at synchrotron radiation source.

1 Owing to the distance between source and specimen, the divergence of the beam normal to the dispersion plane is small and with SR it is unnecessary to perform tilt adjustments to bring the Bragg planes of specimen and beam conditioner exactly parallel.

2 Owing to the small divergence in the dispersion plane the range of wavelengths diffracted from the beam conditioner is always small and dispersion broadening of the SR rocking curve is small, even when the Bragg planes of beam conditioner and specimen are significantly different.

3 As specimen and beam conditioner are both far from the SR source, defects in both are revealed with almost equally good resolution on the topograph.

4 Owing to the very high intensity of SR, exposure times on nuclear plates are typically ten minutes, whereas for such a slow emulsion, exposures are of the order of days in the laboratory. This reduction in timescale is particularly important for multiple exposure topographs of deformed crystals.

A further important consequence of the small source size and low divergence is that it now becomes possible to undertake 'weak beam' topography. Weak beam imaging is a standard technique in electron microscopy. The technique is to set the specimen far from the Bragg peak and then only the region where the lattice planes around the defect are rotated into the Bragg condition gives rise to scattered intensity. As the strain field falls with increasing distance, the diffracting volume falls rapidly and thus the image rapidly narrows. Weak beam topography has not previously been widely practicable because of the large source sizes of earlier synchrotron radiation machines. However, with third-generation sources, the technique works well and it has already been demonstrated at the ESRF.[35] Using simulation code[36] developed for use in the Bragg reflection geometry, the variation in image width with deviation from the Bragg position has been examined. It is seen that the image narrows rapidly out to about twice the perfect crystal reflecting range, but after that decreases rather slowly. The intensity continues to fall and little is gained at large deviations (Figure 10.15).

10.5.5 *Example applications of synchrotron radiation for double-crystal topography*

The earliest application was at LURE in the study of misfit dislocations in GaAlAs epitaxial layers on GaAs[37] using an elegant, fixed wavelength, low divergence beam conditioner designed by Hashizume. Bragg case dislocation images were compared with simulations and excellent agreement was found.[38] Double-crystal topography has the advantage that the substrate and epilayer can be imaged separately and defects in each identified. More recently, at Daresbury, studies of thin InGaAs layers on GaAs have shown how the misfit dislocations are initially nucleated in this highly strained system. This result is of immense technological importance. Substantial relaxation of the strain leaves optoelectronic devices unworkable and thus it is extremely important to understand the formation mechanisms of the misfit dislocations which are responsible for the relaxation. The initial relaxation can be

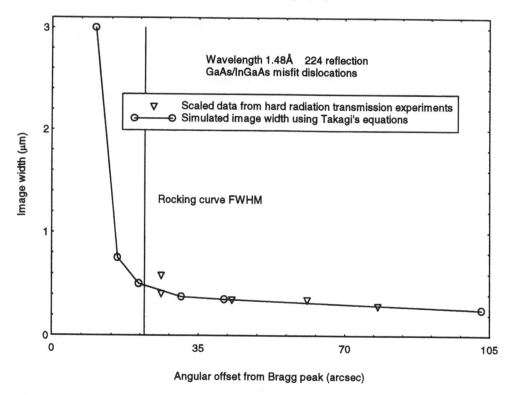

Figure 10.15 Simulated image width as a function of deviation parameter in Bragg case weak beam topographs. Here, the specimen is set off the Bragg peak and an image of the defect occurs only when the lattice planes are locally rotated or dilated back into the Bragg condition. As this occurs only close to the dislocation core, the images are narrowed from those under strong beam conditions

extremely anisotropic, as seen in Figure 10.16. This was believed to be due to the differing mobilities of dislocations on the two types of glide plane in the III–V compounds.

The *in situ* studies have shown, however, that all the misfit dislocations nucleated in the initial stage are not only of 60° type with inclined Burgers vectors, but also of the fast α type.[39] From these studies in very low dislocation density vertical freeze gradient material, it has become apparent that the misfit dislocation anisotropy arises from an anisotropy in the Burgers vector distribution in the standard liquid encapsulated Czochralski crystals. The measurements have given very clear support for the mechanism of misfit dislocation generation proposed by Matthews and Blakeslee.[40] Two critical thicknesses have been observed, one for initial nucleation of misfit dislocations and another, at a substantially greater layer thickness, for misfit dislocation multiplication.

In situ plastic deformation experiments on semiconductors have also been undertaken at LURE by George and colleagues. Tensile loading experiments were performed on sulphur-doped InP[41] in the temperature region of 548 to 648 K. In this range, where InP is ductile, the velocities of screw and 60° dislocations were measured, an activation energy of 1.7 eV being obtained from the temperature-dependence. The velocity of the fast 60° dislocations appeared to depend signifi-

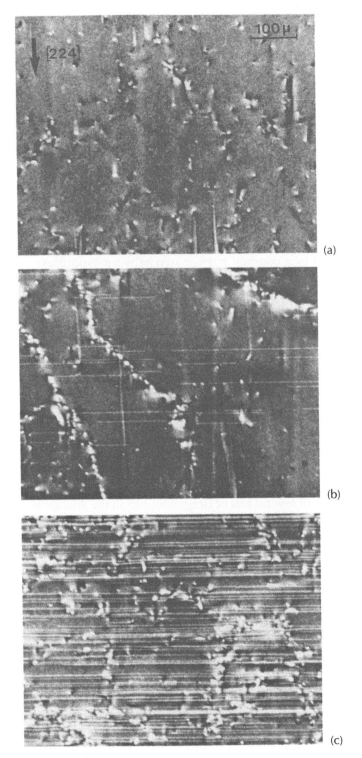

Figure 10.16   Double-crystal topographs of misfit dislocations in InGaAs on GaAs: (a) below, (b) at, and (c) above the critical thickness for relaxation

cantly on the orientation relative to the crystal surface, a feature still not satisfactorily explained. Studies of GaAs and Si have been concerned with the phenomenon of fracture, crack-tip plasticity and the brittle-to-ductile (BDT) transition. Samples were fabricated containing a notch which was then loaded in such a manner that the applied stress intensity factor at the crack tip increased at a constant rate. The BTD transition was found to be extremely sharp,[42] which can be related to the rapid change in dislocation velocities.

An important feature to note in double-axis topography experiments is that when the beam area is large, the measured rocking curve widths are not necessarily intrinsic. For example, mismatched epitaxial layers curve substrate wafers by an amount which depends on the degree of mismatch and layer thickness. Topographs of such curved wafers show bands of diffracted intensity, and it is often possible to image both substrate and epilayer reflections from different parts of the specimen on one plate. These intensity bands can be used to plot the long-range deformation in imperfect crystals by making multiple exposures on a single plate with the specimen moved by a constant angular step between exposures.

This technique, first developed by Renninger[43] using laboratory equipment, has been used to study the relation between deep level, electrically active EL2 defects and lattice perfection in semi-insulating GaAs.[44] Ishikawa et al.[45] performed plane wave topography with a separate monochromator-collimator in the (+,+) setting to produce beautiful 'zebra patterns' of stripes as a different part of the crystal diffracts for successive angular steps. A two-fold symmetric strain was observed in a (001) orientation GaAs wafer with a highly stressed region in In-doped materials owing to impurity segregation. In order to distinguish between tilts and dilations, two sequences of 'zebra-patterns' are required. Provided that the crystal symmetry permits this, the simplest arrangement is to rotate the crystal 180° about its surface normal. Then for one set of contours the tilts and dilations they add, for the other they subtract. It is then possible to determine the separate contours for tilt and dilation. Bilello et al.[46] have used a similar technique with the double-crystal station at the NSLS, Brookhaven, to measure the strains around cracks in fatigued zinc bicrystals. By use of several sets of Bragg contours in the double-axis geometry, they were able to reconstruct the full tensor deformation field around cracks. These experiments prove that high angular sensitivity, but low spatial resolution, X-ray topography has applications outside the confines of highly perfect crystals. Their work is an excellent example of how X-ray topography, with a very high intrinsic strain sensitivity can be applied to heavily deformed materials. X-ray topography is *not* just useful for the study of highly perfect single crystals.

On the other hand, the importance of highly perfect crystals to the semiconductor industry in Japan has meant that considerable effort has been expended in experiments to increase the sensitivity to microstrains. These experiments have all used one or more beam-conditioning crystals to produce a pseudo plane wave, a beam of angular divergence much less than the reflecting range of the specimen crystal. Chikaura, Imai and Ishikawa[47] used an asymmetrically cut reference crystal to give a beam of divergence 0.59 arcsec, less than 0.3 of the range of the specimen reflection, but subsequent experiments have improved on this significantly.[48] As an example of the application of the instrumentation at the Photon Factory, we cite the work of Matsui and colleagues[49] from NEC who have measured the strain fields associated with D type microdefects. Experiments with such highly collimated and low dispersion beams take only a very small fraction of the incident X-ray flux and thus second- or third-generation synchrotron sources are essential if exposure times are to be practicable. We note, however, that the extremely small source size now achieved at the ESRF means that the divergence is always sub-arcsecond at the topography station.

10.5.6 *Triple-axis topography*

In the double-crystal mode, tilts and dilations can only be distinguished by use of more than one reflection and then reconstructing the strain field. A direct method for producing tilt or dilation maps involves the use of an *analyser* crystal after the specimen. In the true triple-axis geometry (Figure 10.17), the analyser selects the scattered intensity from a small volume of reciprocal space. It is easy to see that if the analyser is left fixed and the specimen only is scanned, then rocking curve broadening arises solely from lattice tilts. By recording a topograph while the specimen is scanned over this rocking curve, a band of intensity is recorded on the topograph corresponding to a contour of constant lattice plane *dilation*.[50] Subsequent steps of the analyser permit full contour maps of dilation to be built up by the multiple exposure technique described above. Such experiments are impracticable in the laboratory because of the long exposure times implicit, but are quite feasible at synchrotron radiation sources.

Kitano *et al.*[51] have used the triple-axis topography method to examine the tilts and dilations around growth dislocations in bulk GaAs crystals. A water-cooled silicon 111 reflection was used to remove most of the heat load on the experiment, the following two 553 asymmetric reflections being chosen because the Bragg plane spacing is almost the same as the 008 reflection used for the GaAs specimen. A symmetrically cut silicon crystal was used with the 553 reflection as analyser. Using this arrangement, they were able to measure the lattice spacing variations in regions with different dislocation configurations. Near cell walls, the interplanar spacing contracts, while where dislocations interact with the crystal surface, the interplanar spacing expands. The reasons for these variations are not fully established but believed to be due to different concentrations of point defects.

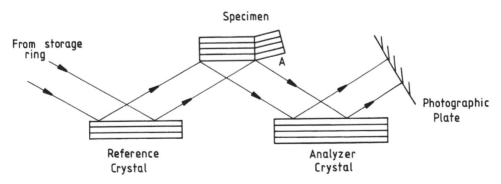

Figure 10.17 Schematic diagram of the triple-axis topography arrangement

10.6 Summary

X-ray topography with synchrotron radiation is now a well established technique for the visualisation and measurement of strains in crystalline materials. Continuing development of storage ring sources is resulting in a steady improvement in capability. The most significant gains in the near future may now well come from attention to detector technology.

References

1. T. TUOMI, K. NAUKKARINEN & P. RABE, Phys. Stat. Sol. (a), **25**, 93 (1974).
2. M. HART, J. Appl. Cryst., **8**, 436 (1975).
3. D. K. BOWEN & B. K. TANNER, Materials Science Reports, **8**, 369 (1992).
4. R. BARRETT, J. BARUCHEL, J. HÄRTWIG & F. ZONTONE, J. Phys. D: Appl. Phys., **28**, A250 (1995).
5. C. REVOL, J. BARUCHEL, D. BELLET, G. MAROT, P. THÉVENEAU & F. ZONTONE, J. Phys. D: Appl. Phys., **28**, A 262 (1995).
6. A. GUINIER & J. TENNEVIN, Acta Cryst., **2**, 133 (1949).
7. L. G. SCHULZ, J. Metals, **6**, 1082 (1954).
8. B. K. TANNER, D. MIDGLEY & M. SAFA, J. Appl. Cryst., **10**, 281 (1977).
9. A. AUTHIER, Adv. X-ray Analysis, **10**, 9 (1967).
10. J. MILTAT & D. K. BOWEN, J. Appl. Cryst., **8**, 657 (1975).
11. J. MILTAT & M. DUDLEY, J. Appl. Cryst., **13**, 555 (1980).
12. S. R. STOCK, H. CHEN & H. BIRNBAUM, Phil. Mag., A, **53**, 73 (1986).
13. W. G. MACHADO, M. MOORE & A. R. LANG, J. Crystal Growth, **71**, 718 (1985).
14. T. TUOMI, M. TILLI & O. ANTTILA, J. Appl. Phys., **57**, 1384 (1985).
15. T. HONDOH, A. GOTO, R. HOSHI, T. ONO, H. ANZAI, R. KAWASE, P. PIMIENTA & S. MAE, Rev. Sci. Inst., **60**, 2494 (1989).
16. S. SUZUKI, M. ANDO, K. HAYAKAWA, O. NITTONO, H. HASHIZUME, S. KISHINO & K. KOHRA, Nucl. Inst. Meths., **227**, 584 (1984).
17. T. HONDOH, H. IWAMATSU & S. MAE, Phil. Mag. A, **62**, 89 (1990).
18. A. HIGASHI, T. FUKUDA, K. HONDOH, K. GOTO & S. AMAKI, in: Dislocations in solids, eds. H. T. SUZUKI, K. NINOMIYA, K. SUMINO & S. TAKEUCHI (University of Tokyo Press, 1985) p. 511.
19. S. AHMAD, M. OHTOMO & R. WHITWORTH, Nature, **319**, 659 (1987).
20. S. AHMAD & R. WHITWORTH, Phil. Mag. A, **57**, 749 (1988).
21. C. SHEARWOOD & R. W. WHITWORTH, J. Glaciology, **35**, 281 (1989).
22. C. SHEARWOOD & R. W. WHITWORTH, Phil. Mag. A, **65**, 85 (1992).
23. B. K. TANNER, Science Progress (Oxford), **67**, 411 (1981).
24. M. SAFA & B. K. TANNER, Phil. Mag. B, **37**, 739 (1978).
25. J. GASTALDI, C. JOURDAN, G. GRANGE & C. L. BAUER, Phys. Stat. Sol. (a), **109**, 403 (1988).
26. J. GASTALDI, C. JOURDAN & G. GRANGE, Phil. Mag. A, **57**, 971 (1988).
27. G. GRANGE, J. GASTALDI, C. JOURDAN & B. BILLIA, J. Crystal Growth, **151**, 192 (1995).
28. J. MILTAT & M. KLEMAN, J. Appl. Phys., **50**, 7695 (1980).
29. R. W. WHATMORE, P. A. GODDARD, B. K. TANNER & G. F. CLARK, Nature, **299**, 44 (1982).
30. C.-C. GLUER, W. GRAEFF & H. MOLLER, Nucl. Inst. Meths., **208**, 701 (1983).
31. H. CERVA & W. GRAEFF, Phys. Stat. Sol. (a), **82**, 35 (1984).
32. H. CERVA & W. GRAEFF, Phys. Stat. Sol. (a), **87**, 507 (1985).
33. A. ZARKA, B. CAPELLE, J. DETAINT & J. SCHWARTZEL, J. Appl. Cryst., **21**, 967 (1988).

34. U. BONSE, in: Direct observations of imperfections in crystals, eds. J. B. NEWKIRK & H. WERNICK (Wiley, New York, 1962) p. 431.
35. F. ZONTONE, Thesis, University of Grenoble (1995).
36. W. SPIRKL, B. K. TANNER, C. WHITEHOUSE, S. J. BARNETT, A. G. CULLIS, A. D. JOHNSON, A. KEIR, B. USHER, G. F. CLARK, C. R. HOGG, B. LUNN & W. HAGSTON, Phil. Mag., A, **69**, 221 (1994).
37. J. F. PETROFF, M. SAUVAGE, P. RIGLET & H. HASHIZUME, Phil. Mag. A, **42**, 319 (1980).
38. P. RIGLET, M. SAUVAGE, J. F. PETROFF & Y. EPELBOIN, Phil. Mag. A, **42**, 339 (1980).
39. S. J. BARNETT, A. M. KEIR, A. G. CULLIS, A. D. JOHNSON, J. JEFFERSON, G. W. SMITH, T. MARTIN, C. R. WHITEHOUSE, G. LACEY, G. F. CLARK, B. K. TANNER, W. SPIRKL, B. LUNN, J. C. H. HOGG, W. E. HAGSTON & C. M. CASTELLI, J. Phys. D: Appl. Phys., **28**, A17 (1995).
40. J. W. MATTHEWS & A. E. BLAKESLEE, J. Crystal Growth, **27**, 118 (1974).
41. A. GEORGE, A. JACQUES & R. COQUILLE, Inst. Phys. Conf. Ser., **76**, 439 (1985).
42. G. MICHOT & A. GEORGE, Inst. Phys. Conf. Ser., **104**, 385 (1989).
43. M. RENNINGER, Z. Angew. Phys., **19**, 20 (1965).
44. G. T. BROWN, M. S. SKOLNICK, G. R. JONES, B. K. TANNER & S. J. BARNETT, in: Semi-insulating III-V materials, eds. D. C. LOOK & J. S. BLAKEMORE (Shiva, Nantwich, 1984) p. 76.
45. T. ISHIKAWA, T. KITANO & J. MATSUI, Japan. J. Appl. Phys., **24**, L968 (1985).
46. J. C. BILELLO, H. A. SCHMITZ & D. DEW-HUGHES, J. Appl. Phys., **65**, 2282 (1989).
47. Y. CHIKAURA, M. IMAI & T. ISHIKAWA, Japan. J. Appl. Phys., **26**, L889(1987).
48. Y. CHIKAURA & M. IMAI, Japan. J. Appl. Phys., **29**, 221 (1990).
49. S. KIMURA, T. ISHIKAWA & J. MATSUI, Phil. Mag. A, **69**, 1179 (1994) .
50. T. ISHIKAWA, T. KITANO & J. MATSUI, J. Appl. Cryst., **20**, 344 (1987).
51. T. KITANO, T. ISHIKAWA & J. MATSUI, Phil. Mag. A, **63**, 95 (1991).

Index